AI For Busy Pastors

101+ Practical Ways To Use AI So You Can Spend More Time Doing What You Are Uniquely Called To Do

Charles C. White

Front and Back Cover design by Anne White

Cover illustration by Gemini AI

Disclaimer

The information, prompts, and guidance provided in this book are for educational and informational purposes only. While the author has made every effort to ensure accuracy and usefulness, the prompts and suggestions are provided "as is" without warranty of any kind, express or implied.

The use of artificial intelligence (AI) tools involves inherent risks and limitations. Users are solely responsible for:

- Verifying the accuracy, appropriateness, and legality of all AI-generated content
- Ensuring compliance with applicable laws, denominational policies, and professional standards
- Exercising pastoral judgment and discernment in all ministry decisions
- Protecting confidential and private information

This book references third-party AI platforms and tools. The author is not affiliated with these companies and makes no representations regarding their reliability, privacy practices, or future availability.

The author and publisher assume no liability for any damages, losses, or consequences arising from the use or misuse of the prompts, techniques, or information contained in this book. This includes but is not limited to theological errors, legal violations, privacy breaches, or pastoral misjudgments resulting from AI-generated content.

Nothing in this book constitutes legal, financial, medical, or professional advice. Readers should consult qualified professionals when such expertise is required.

By using the prompts and methods in this book, you acknowledge that you have read this disclaimer and agree to use the content at your own risk.

If you are not already using AI, but want to test the waters, sign up for the free version of one or more of these platforms (current as of early 2026). For more options, see the last page of the book.

Claude (Anthropic)   **claude.ai**
Known for longer context windows and nuanced conversations, particularly strong for writing complex documents.

ChatGPT (OpenAI)   **chat.openai.com**
Excellent for writing tasks across all ministry areas.

Google Gemini   **gemini.google.com**
Integrated with Google Workspace, making it seamless for pastors and churches already using Gmail, Google Docs, and Google Drive.

Perplexity   **perplexity.ai**
Research-focused AI that cites sources, excellent for fact-checking, research, and finding scholarly resources quickly.

This book focuses on specifically how to *use* AI.

If you want to read more about the history, overarching concepts, and social / ethical concerns of AI, I highly recommend Jason Moore's book *AI and the Church: A Clear Guide for the Curious and Courageous*. Here's his description of the book:

"The world is rapidly changing, and artificial intelligence (AI) is at the forefront of that transformation. But what does this mean for the church? AI and the Church: A Clear Guide for the Curious and Courageous explores the intersection of faith and technology, providing a thoughtful, balanced, and practical approach to integrating AI into ministry.

"Whether you're a church leader, tech enthusiast, or someone simply curious about AI's impact on society, this book offers a clear roadmap for understanding and utilizing AI tools. From chatbots and image generation to AI-driven analytics and video creation, AI and the Church equips you with the knowledge and confidence to embrace AI in a way that aligns with your mission.

"Jason Moore – an author, speaker, and seasoned expert at the nexus of technology and ministry – demystifies AI, addressing common fears and ethical concerns while offering practical applications for enhancing worship experiences, outreach efforts, and administrative tasks. This is not just a tech manual; it's a thoughtful guide that considers the theological implications of AI and how it can be used to further the Great Commission.

"Inside, you'll discover:
- How AI is already part of your daily life, even in ways you might not realize
- Ten practical ways to use AI in the church today
- Guidance on navigating the ethical and spiritual questions AI raises
- Tools to enhance your church's ministry, regardless of your congregation's size
- Steps to develop an AI policy for your church or organization

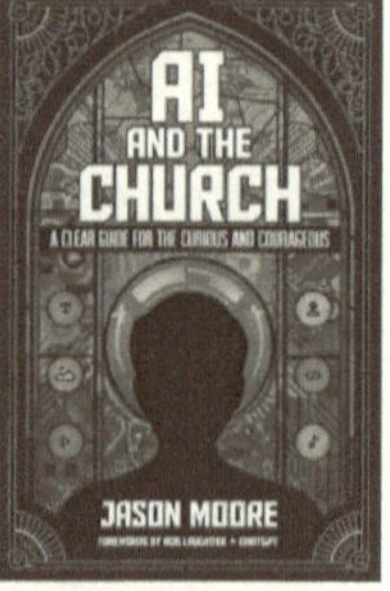

"AI and the Church is a must-read for anyone looking to understand AI's role in the future of ministry and how to harness its potential for good. Join the conversation and discover how AI can help the church not only survive but thrive in the digital age."

Table of Contents

CHAPTER 0: BEFORE YOU GET STARTED

SECTION 1: SERMON & WORSHIP PREPARATION

SECTION 2: THE LITURGICAL CALENDAR & SPECIAL DAYS

SECTION 5: EDUCATION & DISCIPLESHIP

SECTION 6: ADMINISTRATION & LEADERSHIP

SECTION 7: CREATIVE & MEDIA

INTRODUCTION
Please read this! ☺

*AI can responsibly handle many of the time-intensive and administrative tasks of ministry – freeing pastors, staff, and volunteers to **focus on what's most important:** human presence, spiritual discernment, and compassionate leadership.*

AI Won't Replace Pastors – But It Can Help Us Be Better Ones

I didn't go to seminary to become a content factory. I went because I felt called to help people connect with a loving God so that they can better love themselves and each other. I wanted to preach, teach, care for souls, and build communities that love God and others.

I was trained to celebrate with people in their joy, walk with them through their grief, help them wrestle with Scripture and their faith, and encourage them to notice God's presence in their lives. But here's what I didn't anticipate: the sheer *volume* of words a pastor must produce every single week.

According to the United Methodist Church's *Book of Discipline 2020-2024* – our denomination's constitution and law, i.e. our "rulebook" – pastors are responsible for at least 39 distinct, ongoing tasks (¶ 340). We preach and teach. We counsel and visit. We marry and bury. We administer sacraments, oversee budgets, recruit leaders, write reports, manage facilities, ensure compliance with denominational policies, model stewardship, lead strategic planning, participate in community partnerships, and attend conference events. Other denominations have similarly broad and staggering job descriptions.

We're *also* expected to be administrators, event planners, fundraisers, social media managers, graphic designers, and ambassadors in our communities – often all in the same day. If you're a pastor reading this, you know *exactly* what I'm talking about.

A typical day might look like this: hospital visit with a dying parishioner at 9am, community food bank board meeting at 11am, lunch with a concerned church member, finance committee budget review at 2pm, followed by prepping for and leading an evening Bible study on the Gospel of Luke.

And in between each of those, you do your best to return emails, phone calls, text messages, DMs, and post something relevant and interesting to the church's social media account(s). Whew!

Each of those activities requires a completely different emotional register, skill set, and energy level. No wonder we're exhausted! No wonder our brains run out of words even though the next sermon, prayer, newsletter article, small group questions, or social media post is due. No one person can be great at everything, and pastors are no exception.

Once I began using AI, I quickly realized that, **when used as my personal Executive and Research Assistant, AI can help me carry the load of pastoral responsibilities** – creating content, managing operations, and planning tasks. **This frees my time and energy for the tasks that requires irreplaceable human pastoral presence:** sitting with families in grief, navigating church conflict with wisdom and grace, spiritual formation that happens in relationship, and the incarnational ministry of simply being present.

AI won't visit your congregant in the ICU. AI won't discern the unspoken pain in someone's prayer request. AI won't know when to sit in silence during a counseling session or when your leadership team needs encouragement or refocusing rather than data analysis.

But AI **can**:

- Compile research on sermon topics in minutes instead of hours
- Draft policy documents you've been "meaning to write" for weeks… or years
- Turn your written sermon (or just your outline) into a blog post, small-group guide, and three social media posts
- Create volunteer tasks and schedules that respect people's time and availability
- Generate first-draft prayers, emails, and newsletters that you then personalize with your voice

My Background: From Printing to Pulpits

I spent 25 years in the printing and publishing industry before I went to seminary. I managed production schedules, coordinated client projects, handled sales and marketing, and learned how to produce high-quality content efficiently and on deadlines.

I earned my Master of Divinity from Claremont School of Theology in 2020, worked as the Office Admin at Sparks UMC in Sparks, NV, and now serve United Methodist Churches in the California-Nevada Annual Conference. As a pastor, I brought that business and production mindset with me.

I'm not a "tech person" who loves AI for its own sake. I'm a pragmatist who understands workflows, deadlines, and deliverables. When ChatGPT was publicly released in late 2022, I immediately recognized it as a **production assistant** for the content-heavy side of ministry.

My production background taught me about templates, batch processing, and repurposing content. AI can do the same thing for pastoral work: it can help you create reusable frameworks, streamline repetitive tasks, and multiply the impact of your best content.

This Book is Built on Three Core Benefits I've Experienced:
1. **AI helps me spend less time *doing* tasks so I can spend more time *being* present** with people in my congregation and community.
2. **AI helps me do more research and create better content** than I could produce alone when my body, mind, and soul are running on fumes.
3. **AI helps me accomplish tasks outside my skillset** – I have no artistic ability, but AI helps me create visual content I can only describe in words.

My goal is ***not*** AI dependence. My goal is **AI-assisted discernment** that helps us become more effective pastors. AI can help us, not replace us.

I'm writing this book from my perspective as a pastor in the United Methodist Church serving moderate and progressive congregations in the incredibly diverse state of California. My prompts reflect Wesleyan theology and my local contexts. **You'll need to adapt them for YOUR context.** If you're Presbyterian, swap in Reformed theology. Greek Orthodox? Ditto. If your congregation skews younger, adjust the tone. If you serve in a rural context, add local details.

That's the beauty of AI: it's an infinitely adaptable tool. The more you tell it *exactly* what you need, the better it can help you.

A Word About AI and Pastoral Integrity

Some pastors worry that using AI is "cheating" or inauthentic. I understand that concern, but I don't share it. When I use a concordance to find cross-references, I'm using a tool. When I consult a commentary to understand historical context or theological perspectives, I'm using a tool. When I use a thesaurus to find the best word, I'm using a tool. When I use my laptop's spelling and grammar checkers, I'm using tools. AI is a tool – a *very* powerful one – but still a tool.

The sermon is still mine. The pastoral care is still mine. The discernment is still mine. AI just helps me offload the labor for the parts that don't require my unique pastoral presence so that I can focus on the parts that only I can do.

AI is not cheating. AI is a tool and we are the stewards of the gifts and time God has given us.

Let's Get Started

Ministry is hard. You're doing sacred work in a world that often doesn't understand or value it. You're stretched thin, often underpaid, and often expected to be excellent at everything.
AI won't solve all your problems. But it can give you back several hours each week **so you can be the pastor your people need.**
And *that* is worth learning a new tool. Let's dive in!

For a PDF of this book, email me at:
`AIForChurch@pm.me`

CHAPTER 0
Before You Get Started

Welcome to your AI toolkit for ministry. Before you dive into the specific prompts in this book, let's establish some foundations that will make every chapter more effective. Think of this as the instruction manual that makes all the tools work better.

Think about how AI can enhance your team effectiveness by asking yourself, "Which member of my team could use this prompt to lighten their workload?" You don't have to be the only person using AI. Your church's worship leader, youth pastor, administrative assistant, or volunteer coordinator might also benefit from these tools. Think beyond yourself – how can AI multiply your team's effectiveness?

Universal Questions to Ask Yourself
Before you use any prompt in this book, ask yourself these four questions. They'll help you customize AI's output to fit your actual ministry context:

1.   What theological distinctions of my tradition should I add to this prompt?
AI doesn't know you're Wesleyan, Reformed, Pentecostal, or Baptist unless you tell it. A prompt that works beautifully for a progressive UMC congregation might need significant adjustment for a conservative Reformed church. Always specify your theological perspective when it matters.

2.   How would I adapt this for my specific community demographics?
A sermon illustration that lands with retirees might fall flat with young families. Social media content that works in urban contexts might miss the mark in rural communities. Always customize your prompts for the people you actually serve.

3.   What would make this output more authentic to my voice and ministry?
AI can generate content that sounds like "a pastor." You can even tell it to write a sermon that sounds like your favorite famous pastor – but please only do that to humor yourself. Your job is to make it sound like

you – *your* sense of humor, *your* pastoral heart, *your* way of explaining Scripture, *your* local references. The more you add *your* voice, the more useful the output becomes.

Your Ministry Context / Identity Profile

Here's a time-saving strategy: create a reusable "context template" that you can copy-paste into most prompts. Fill this out once, save it somewhere accessible (a note on your phone, a document on your desktop), and use it over and over.

Template:

```
I'm a [denomination] pastor serving a [size]-member
congregation in [location]. Our congregation is
primarily [age demographics and any other relevant
details]. My theological perspective is
[Wesleyan/Reformed/Orthodox/Pentecostal/etc.].
```

My example:

```
I'm a United Methodist pastor serving a 120-member
congregation in Sacramento, CA. Our congregation is
primarily 65+ years old with a progressive
theological outlook. My theological perspective is
Wesleyan.
```

Your version will be unique to you. If you're a Presbyterian pastor in suburban Atlanta serving young families, say that. If you're a Baptist pastor in rural Montana with a traditional congregation, say that. The more specific you are, the better AI can tailor its responses.

✅ **Pro tip:** Save multiple versions if you wear multiple hats. You might need one template for preaching, another for your community board work, and another for denominational responsibilities.

Standard Parameters for Your Context

Beyond theological perspective and demographics, you'll often want to include specific technical parameters like length, tone, reading level, and formatting. Here's a template for those details:

Parameter Template:

```
Length: [word count or page equivalent]
Audience: [description of who will read/hear this]
```

```
Reading level: [grade level - most church content
works well at 9th-11th grade]
Tone: [warm, formal, conversational, prophetic, etc.]
Style: [1st or 3rd person, formal, humorous, etc.]
Special formatting: [capitalization, punctuation,
structure preferences]
```

My example parameters for most content:
```
Length: 200-250 words
Audience: Educated adults of all ages in a 120-member
United Methodist congregation (primarily 65+ years
old)
Reading level: 11th grade (simple, clear language
with minimal religious jargon)
Tone: Warm, encouraging, and thoughtful
Special formatting: Capitalize the first letter of
divine pronouns (He, Him, His, etc.)
Style: Avoid overly academic language or complex
theological terminology; use EN dashes ( - ) instead
of EM dashes
```

Customize this for your context. If you're writing for youth, drop the reading level to 7th-8th grade and make the tone more casual. If you're writing formal denominational reports, adjust accordingly.

Create multiple chat sessions for different purposes. While the prompt above is structured and authoritative, some pastors find a more "Creative Partner" approach better for brainstorming. This version focuses less on strict rules and more on the process of discovery.

Template for a Creative Partner
```
"I want you to act as a Theological Research
Assistant. Your goal is to help me 'think out loud.'
When I ask a question, don't just give me one answer.
Instead, provide three distinct perspectives:
- The Traditional View: How would a classical
theologian or a historic creed approach this?
- The Contemporary View: How does this intersect with
modern culture or current events?
- The 'Odd' View: Give me a creative, unexpected, or
counter-intuitive angle that might spark a unique
discussion in Bible Study or sermon illustration.
Keep your responses concise and always end with a
follow-up question that challenges me to dig deeper
into the pastoral implications of the topic."
```

By establishing these parameters early, the AI moves from being a generic search engine to a specialized "production assistant" that understands the nature of *your* denomination, *your* theology, and *your* specific needs for *your* church and *your* community. Using these detailed prompts at the beginning of chat sessions prevents the need to constantly remind the AI to "keep it Wesleyan" or "don't use the King James Version."

Garbage In, Garbage Out: Why Details Matter
Here's the most important principle in this entire book: **The better your prompt, the better your result.** AI is not a mind reader. It doesn't know what you're thinking, what you want, or what you're trying to avoid unless you tell it explicitly. Vague prompts produce vague results. Detailed prompts produce useful results.

Bad prompt example: `"Write a sermon on grace."`
It's bad because it's **too vague.** What Scripture passage? What angle on grace? What theological tradition? What audience? AI will give you something generic and probably unusable.

Good prompt example:
`"I'm a United Methodist pastor preparing a sermon on Ephesians 2:8-10 for a congregation of primarily older adults. Write a sermon outline that emphasizes grace as God's free gift while also addressing the role of good works in the Christian life. My theological perspective is Wesleyan. Include 3 main points, sub-points, and potential illustrations. Length: 300-400 words."`

This is much better because it defines a specific Scripture, clear theological angle, defined audience, stated perspective, structural expectations. AI now has enough details to give you something useful.

The magic end to many prompts:
Once you've entered your prompt and attached any relevant documents, end your prompt with this question:
`"Before you get started, do you have any questions or need any additional information from me?"`

You may be amazed at the clarifying questions your AI asks you. Answering these questions at the beginning often creates dramatically better (more detailed or concise, more accurate, more relevant) results.

The magic follow-up prompt:

When AI gives you something that misses the mark, don't just try again with a new prompt. Instead, tell AI how it missed the mark and how you want it to adjust to give a better result. Continue this back-and-forth exchange until you get what you want. At the end of the conversation, ask:

```
"What's a better way to word my original prompt that
would have produced the result I actually wanted?"
```

This teaches you AND the AI. By customizing AI's suggestions, you'll learn how to write better prompts. As you do, AI will better understand what you're actually looking for. It's collaborative learning.

Do You Need a Paid Account?

Most AI platforms offer both free and paid versions. Here's my honest assessment: **Start with free accounts.** For most pastoral tasks in this book, free versions of ChatGPT, Claude, or other platforms will work fine. You can write sermons, draft newsletters, create discussion questions, and repurpose content without spending a dime.

Consider upgrading if:
- You regularly hit usage limits during busy seasons (Advent, Lent, Easter, VBS planning)
- You need faster response times or priority access
- You're working on complex, multi-step projects that require lots of back-and-forth
- You want access to advanced features like web search, image generation, or longer context windows

Strategic upgrade timing:

Pay monthly during your busiest seasons rather than committing to annual subscriptions. Upgrade in November for Advent/Christmas planning, downgrade in January. Upgrade in February for Lent/Easter, downgrade in May. You'll save money and still have access when you need it most.

Team sharing:
If your church has the budget for it, one paid account shared strategically among staff can be more cost-effective than multiple free accounts. Just be mindful of usage limits and privacy – don't share login credentials carelessly.

Bottom line: Don't let cost be a barrier. Free accounts are powerful enough for 90% of what's in this book. Upgrade only when you're consistently bumping into limitations.

The Three Critical Habits

No matter which prompts you're using or which chapter you're working from, always practice these three habits:

Habit 1: Always verify Scripture references and theological claims.

AI is not a theologian. It sometimes cites Bible verses that don't exist, misapplies passages, or suggests theological connections that don't hold up under scrutiny. Before you use any Scripture reference or theological claim AI generates, look it up. Check the context. Make sure it actually says what AI thinks it says.

Habit 2: Always adapt the output to your voice.

AI writes like "a pastor" – generic, broadly acceptable, theologically safe. You need to sound like you. Add *your* stories, *your* sense of humor, *your* pastoral heart, *your* way of explaining things. The goal is **not** to use AI's words verbatim; the goal is to use *your* ideas as a starting point and use AI to help you structure and develop *your* ideas in your own pastoral voice.

Habit 3: Always consider your specific congregation's needs.

AI doesn't know *your* people. It doesn't know that your congregation just experienced a tragedy, that your community is dealing with economic hardship, that your church is in the middle of a building campaign, or that your youth group is small and thriving right now. You know those things. Always ask yourself: "How does this content need to change based on what my congregation is actually experiencing?"

How to Read This Book

This book contains practical prompts organized into ministry areas:

- **Sermon & Worship Preparation** – outlines, illustrations, prayers, liturgy, and special services
- **Communication & Outreach** – newsletters, social media, emails, and website content
- **Pastoral Care & Counseling** – condolence letters, crisis response, and care coordination
- **Education & Discipleship** – small group guides, Bible studies, and teaching materials
- **Administration & Leadership** – policies, reports, planning documents, and organizational tools
- **Creative & Media** – repurposing content, graphics, and multimedia projects
- **Research & Professional Development** – theological research, data analysis, and continuing education

You don't need to read this book cover-to-cover. Each chapter stands alone. Jump to the chapter that addresses your most urgent need today. Use the prompt, customize it for your context, and let AI handle the first draft so you can focus on making the results exactly what **you** need.

The structure of every chapter:
1. **Our Goal** – what we're trying to accomplish with AI
2. **The Problem** – why this task is difficult or time-consuming
3. **Why This Matters** – the significance to ministry
4. **The Prompt** – copy-paste ready, with customization points in [brackets]
5. **What to Expect** – brief description of AI's output
6. **Follow-Up Prompts** – ways to refine, expand, or redirect AI's response
7. **Warnings** – when relevant (Scripture accuracy, privacy, discernment reminders)
8. **How to Use It** – practical next steps for using the AI output
9. **Reflection Question** – helps you apply this to your specific context

The warnings are important!
Throughout this book, you'll notice the following warning boxes attached to many chapters. These aren't just cautionary decorations – they're essential guardrails. AI is a powerful tool, but it's not infallible. It can generate Scripture references that don't exist, suggest legal advice that doesn't comply with your state's laws, propose pastoral responses

that miss crucial cultural or relational dynamics, or create theologically plausible-sounding content that does not match your personal or denominational beliefs.

📖 **Scripture Check** – verify Biblical accuracy
⚠ **Verify This** – fact-check claims/sources
🙂 **Your Discernment Matters** – AI can't replace pastoral judgment
🔒 **Privacy Reminder** – protect confidential information
✝ **Theological Caution** – denominational/doctrinal nuance needed

The warnings remind you where AI is most likely to get things wrong and where your pastoral judgment, theological training, and critical thinking are essential. Think of AI as a very enthusiastic intern who's read everything but doesn't have wisdom, discernment, or accountability. Your job isn't to trust AI blindly – it's to use AI strategically while bringing your pastoral expertise, contextual knowledge, and relationship with God to every decision. The warnings show you where to pay closest attention.

Learning Principles, Not Just Prompts
The patterns you'll learn in this book are transferable. When you see a follow-up prompt like "Make this shorter" or "Rewrite this for a different audience," **you're learning concepts and skills** that work across all AI platforms and all types of content. The specific prompts in this book may eventually become outdated as AI technology evolves, but the *principles* behind them will remain valuable.

Remember: Start with your Ministry Context / Identity Profile "Standard Parameters" previously discussed and build from there.

Now that we've covered "theory," let's start putting these ideas to good use!

SECTION 1

SERMON & WORSHIP PREPARATION

CHAPTER 1
Sermon Research and Exegetical Background

Our Goal
Quickly gather historical context, cultural background, theological insights, and cross-references for a Scripture passage without spending hours in commentaries.

The Problem
Solid Biblical preaching requires research – understanding the original language, historical setting, cultural context, and how the passage fits into the broader biblical narrative. But research is time-consuming. By the time you've consulted three commentaries, tracked down cross-references, and explored the Greek or Hebrew, hours have passed and you haven't started writing.

Why This Matters
Good exegesis is the foundation of faithful preaching. When you understand what the text meant in its original context, you're better equipped to apply it faithfully today. AI can accelerate the research phase, giving you a solid foundation in minutes so you can spend your time on pastoral application and sermon craft.

The Prompt
```
I'm a [denomination] pastor preparing to preach on
[Scripture passage]. My theological perspective is
[Wesleyan/Reformed/Orthodox/etc.].

Provide exegetical background including:
- Historical and cultural context of the passage
- Key themes and theological insights
- Important word studies (Greek/Hebrew if applicable)
- How this passage fits into the broader Biblical
narrative
- 3-5 relevant cross-references with brief
explanations
- Potential interpretive challenges or debates

Audience: Pastor preparing a sermon
Length: 400-500 words
Style: Scholarly but accessible
```

What to Expect
AI will generate comprehensive background covering historical setting, cultural context, key themes, word studies, and cross-references. The information will be generally accurate but **should be verified** against trusted commentaries.

Follow-Up Prompts
"Expand the section on [specific theme or word] with more detail."
This lets you dive deeper into one aspect without re-reading all the background material.
"What are the main interpretive differences between Wesleyan and Reformed theologians on this passage?"
You can explore how your theological tradition specifically approaches a text.
"Suggest three contemporary issues or questions this passage addresses."
This helps you bridge from ancient context to modern application.

📖 **Scripture Check**: Always verify AI's exegetical claims against trusted commentaries and Biblical scholarship. AI sometimes oversimplifies complex interpretive issues or misses important textual nuances.

✅ **Pro tip:** Ask AI "What translations of the Bible do you have direct access to?" You can then tell it which translation you prefer OR you can copy-paste the translation you've chosen.

How to Use It
1. Use AI's research as your starting framework and verify key claims in trusted commentaries
2. Ask AI to expand specific sections (word studies, cultural context) that need more depth
3. Have AI generate contemporary application ideas that bridge from ancient context to modern life

Reflection Question
Which insight from this research most challenges or reshapes your initial understanding of the passage, and how will that affect your sermon's direction?

CHAPTER 2
Creating Sermon Outlines from Scripture Passages

Our Goal
Generate a structured sermon outline based on a specific Scripture passage that includes main points, sub-points, and potential illustrations.

The Problem
Staring at a blank page (or screen) can be exhausting when you're trying to craft a sermon. You know the passage, you've prayed over it, but organizing your thoughts into a coherent outline can take hours – especially when you're juggling hospital visits, meetings, and pastoral care.

Why This Matters
A solid outline is the foundation of an effective sermon. It helps you stay focused, ensures theological coherence, and makes the writing process much faster. When AI handles the initial structure, you can spend your energy refining the theology and adding your pastoral voice rather than wrestling with organization.

The Prompt
```
I'm a [denomination] pastor preparing a sermon on
[Scripture passage]. My congregation is [description:
size, demographics, context]. My theological
perspective is [Wesleyan/Reformed/Orthodox/etc.].
My theme for this sermon is [faith/hope/love/etc.].

Create a sermon outline with:
- A compelling sermon title
- 3-4 main points drawn from the passage
- 2-3 sub-points under each main point
- Suggested transitions between points
- Potential illustration ideas for each main point
- A strong conclusion with a call to action

Length: Outline format, approximately 300-400 words
Style: Clear, theologically sound, accessible to
educated adults
```

What to Expect
AI will generate a complete outline with hierarchical structure (main points → sub-points), transition suggestions, and illustration ideas. The

theology will be generally sound but may need your denominational refinement.

Follow-Up Prompts
"Expand main point #2 with more biblical cross-references."
This helps you add depth to one section without rewriting the entire outline – useful when one point needs more theological grounding.
"Adjust this outline for a youth audience instead of adults."
Once you have a solid outline, you can ask AI to rewrite it for different age groups or contexts without starting from scratch.
"Give me three different sermon titles for this outline."
Generating multiple options helps you pick the title that best fits your congregation and the moment.

📖 **Scripture Check**: Always verify that cross-references actually support the points being made. AI sometimes suggests passages that sound thematically related but don't hold up exegetically.

✅ **Pro tip:** If your word-count is critical, ask AI "How do you count words?" You can then tell it to ignore headlines and sub-heads OR tell it to count every single word it writes.

How to Use It
1. Use AI's outline as your foundation and verify all Scripture references for accuracy
2. Ask AI to expand specific main points that need more Biblical cross-references or depth
3. Have AI adjust the outline for different audiences (youth, adults, seekers) by requesting variations

Reflection Question
What local context or recent events in your congregation or community should you add that AI couldn't know?

CHAPTER 3
Generating Sermon Illustrations and Stories

Our Goal
Create compelling illustrations that illuminate Scripture and connect Biblical truth to everyday life without spending hours searching illustration databases or your memory.

The Problem
Great illustrations make abstract biblical concepts concrete and memorable. But finding the right story – one that fits your point, resonates with your congregation, and doesn't feel clichéd – can take longer than writing the sermon itself. You need fresh, relevant illustrations, and you need them now.

Why This Matters
Illustrations are bridges between ancient text and modern life. They help people see how Scripture intersects with their actual experiences. When AI helps you generate illustration ideas quickly, you can spend more time refining and personalizing them rather than hunting for something usable.

The Prompt
```
I'm preaching on [Scripture passage] with the main
point: [your key theological claim or application].

Generate 3-5 sermon illustration ideas that:
- Connect this biblical truth to contemporary life
- Are relatable to [audience description: e.g., older
adults, young families, mixed-age congregation]
- Avoid clichés or overused stories
- Include a mix of types: personal story concepts,
current events, everyday observations, historical
examples

Context: [Brief description of your congregation and
community]
Tone: Accessible, memorable, authentic
Length: 50-75 words per illustration idea
```

What to Expect
AI will generate multiple illustration concepts with brief descriptions. They'll be starting points – you'll need to flesh them out, personalize them, or use them to trigger your own memories and ideas.

Follow-Up Prompts
"Expand illustration #2 into a full 150-word story."
This helps you develop the most promising idea into sermon-ready form.
"Make illustration #3 more specific to a rural/urban/suburban context."
You can adapt generic ideas to match your actual community.
"Suggest an illustration that addresses potential objections or doubts about this point."
Sometimes the best illustrations acknowledge struggle rather than only celebrating victory.

Your Discernment Matters: AI-generated illustrations can feel generic or disconnected from real life. Always personalize them with your own observations, experiences, or stories from your congregation. The best illustrations come from your actual pastoral presence in people's lives.

How to Use It
1. Review all the illustration ideas AI generates
2. Pick the 1-2 that resonate most with your congregation's experience
3. Personalize your illustration with local details, names (with permission), or your own observations
4. Consider replacing AI's suggestion entirely if it triggers a better idea from your own life

Reflection Question
Which of AI's illustration ideas connects most naturally to a real conversation, observation, or experience you've had recently with someone in your congregation, and how can you use that real-life connection instead of AI's generic version?

CHAPTER 4
Crafting Sermon Introductions and Hooks

Our Goal
Create compelling sermon openings that grab attention, establish relevance, and draw listeners into the Biblical text.

The Problem
You have seven seconds to capture your congregation's attention before minds start wandering. A weak introduction – starting with "Today we're going to talk about..." or diving straight into the Biblical text without context – loses people before you've even begun. But crafting a strong hook takes creative energy you might not have after writing the rest of the sermon.

Why This Matters
A great introduction does three things: it captures attention, it establishes why this sermon matters today, and it creates curiosity about where you're headed. When AI helps you generate multiple opening ideas, you can choose the one that best fits your congregation and the moment.

The Prompt
```
I'm preaching a sermon on [Scripture passage] with
the title "[sermon title]" and the main idea: [brief
summary of your key point].

Create 3 different sermon introduction options (100-
150 words each) that:
- Open with a compelling hook (question, story,
observation, or relevant cultural moment)
- Establish why this sermon matters to [audience
description]
- Create curiosity without giving away the whole
sermon
- Transition naturally into the Scripture passage

Tone: Engaging, conversational, relevant
Avoid: Clichéd openings, guilt-inducing questions, or
overly academic approaches
```

What to Expect
AI will generate three distinct sermon opening approaches – likely one story-based, one question-based, and one observation-based. Each will be 100-150 words and ready to adapt.

Follow-Up Prompts
`"Make introduction #1 more personal and vulnerable."`
This adjusts the tone to create more pastoral intimacy and authenticity.
`"Rewrite introduction #2 to reference a recent news event or cultural moment."`
You can make the opening feel more timely and relevant.
`"Create a fourth option that opens with a surprising statistic or fact."`
Sometimes you need a different angle than the first three options provided.

How to Use It
1. Read all three introductions out loud to hear how they sound
2. Pick the one that feels most authentic to your voice
3. Personalize with a real story, local detail, or current event
4. Make sure the transition to Scripture feels natural, not forced

Reflection Question
Which of the AI-generated introductions would most effectively reach the person in your congregation who's skeptical about church, exhausted by life, or distracted by worry, and how can you refine it to speak directly to them?

CHAPTER 5
Applying Scripture to Current Events and Cultural Moments

Our Goal
Connect Biblical truth to contemporary issues, cultural conversations, or current events in ways that feel relevant without being partisan or reactive.

The Problem
Your congregation lives in the real world. They're processing news, cultural shifts, political tensions, and social movements. When you preach Scripture as if these things don't exist, you risk making the Bible feel disconnected from daily life. But wading into current events can be pastorally risky – you don't want to alienate people or make the sermon feel like political commentary.

Why This Matters
Faithful preaching shows how Scripture speaks to the world we actually inhabit. When you carefully connect Biblical truth to contemporary issues, you demonstrate that God's word is living and active, not locked in the past. AI can help you explore those connections thoughtfully without rushing into reactive, poorly considered applications.

The Prompt
```
I'm preaching on [Scripture passage] and want to
apply it to [current event, cultural trend, or social
issue].

Help me explore how this passage speaks to this
contemporary moment:
- Identify Biblical principles connecting to this
issue
- Suggest thoughtful, pastorally sensitive ways to
make the connection
- Highlight potential pitfalls or misapplications
- Offer language that invites reflection rather than
demanding agreement

My congregation: [description: political diversity,
theological perspective, cultural context]
Theological tradition: [Wesleyan/Reformed/etc.]
Tone: Thoughtful, humble, pastorally wise
Length: 200-300 words
```

What to Expect
AI will suggest ways the Biblical text connects to the contemporary issue, offer pastoral framing, and flag potential sensitivities. The suggestions will be cautious and balanced – you'll need to decide how directly to engage based on your congregational context.

Follow-Up Prompts
"How might someone in my congregation who holds [specific viewpoint] receive this application?"
This helps you anticipate reactions and consider whether your framing creates space for diverse perspectives.
"Suggest a way to apply this passage to the issue without explicitly naming the political debate."
Sometimes the most faithful approach addresses underlying values without wading into partisan territory.
"Offer a pastoral caveat acknowledging the complexity of this issue."
This helps you preach with conviction while maintaining humility.

📖 **Scripture Check**: Make sure the Biblical text actually addresses the issue you're connecting it to. Avoid forcing connections that aren't exegetically sound.

🙂 **Your Discernment Matters**: AI can't know your congregation's political diversity, recent conflicts, or tender spots. Use extreme pastoral care when applying Scripture to divisive issues.

How to Use It
1. Review AI's suggestions for theological soundness
2. Consider your congregation's readiness for this application
3. Consult a trusted colleague before preaching on highly sensitive topics

Reflection Question
Is this the right time and context for this application, or would it be more faithful to address this issue in a setting where there's space for dialogue?

CHAPTER 6
Developing Children's Messages from Adult Sermon Themes

Our Goal
Transform your adult sermon content into age-appropriate children's messages that teach the same Biblical truth in ways kids can understand and remember.

The Problem
You've spent hours preparing your sermon for adults, and now you need a children's message that connects to the same theme – but talks about covenant theology or justification by grace aren't going to land with six-year-olds. Creating an entirely separate message for kids feels like starting from scratch, and you're out of time and creative energy.

Why This Matters
Children's messages are formative moments in a child's faith journey. When your children's message connects to your adult sermon, you're reinforcing the same Biblical truth across generations and helping families discuss what they learned together. AI can help you translate complex theology into concrete, age-appropriate stories and object lessons without doubling your prep time.

The Prompt
```
I'm preaching on [Scripture passage] with the main
theme: [brief summary of your key theological point].

Create a children's message (3-5 minutes when spoken)
that:
- Teaches the same Biblical truth in age-appropriate
language
- Uses a concrete object lesson, simple story, or
interactive activity
- Connects to children's everyday experiences
(school, family, friends, play)
- Includes 2-3 questions to engage kids in
conversation
- Ends with a simple, memorable takeaway

Target age: [e.g., 5-10 years old, elementary age]
Tone: Warm, playful, engaging
Avoid: Abstract theological language, guilt-inducing
questions, or talking down to children
Length: 200-250 words
```

What to Expect

AI will generate a complete children's message with an object lesson or story, discussion questions, and a clear takeaway. The language will be simplified, and the concepts will be made concrete through everyday examples kids can relate to.

Follow-Up Prompts

```
"Make this more interactive – add a physical activity
or movement the kids can do."
```
This helps you create participation and keeps kids engaged rather than just listening passively.
```
"Suggest three different object lessons I could use
to teach this same truth."
```
You can pick the object lesson that uses materials you already have on hand.
```
"Adjust this for teenagers instead of elementary
children."
```
Youth need different examples and less simplification – AI can adapt the same truth for older kids.

How to Use It

1. Read the children's message out loud to test timing and flow
2. Ask AI to suggest different object lessons if you don't have the materials on hand
3. Have AI adjust the message for different age ranges (preschool vs. elementary)
4. Be ready to improvise – kids are unpredictable, and that's okay

Reflection Question

What recent experience or conversation with a child in your congregation could you weave into this message to make it feel personally connected to their lives?

CHAPTER 7
Writing Calls to Worship

Our Goal
Create compelling calls to worship that set the tone for the service, invite the congregation into God's presence, and connect to the day's theme or Scripture.

The Problem
The call to worship is the doorway into corporate worship, but writing fresh, meaningful liturgy every week can feel like one more creative task piled onto an already overwhelming sermon prep load. You want something better than recycling the same generic phrases, but you don't have hours to craft original liturgy.

Why This Matters
A strong call to worship shifts people from "getting settled in the pew" mode to "encountering the living God" mode. It establishes the theme, invites participation, and reminds the congregation why we've gathered. When AI helps you draft calls to worship quickly, you can create liturgy that feels fresh and connected to your sermon without starting from scratch every week.

The Prompt
```
I'm planning worship for [season/Sunday: e.g., Third
Sunday of Lent, Ordinary Time, Easter Sunday] based
on [Scripture passage] with the theme of [brief theme
description].

Create a call to worship that:
- Opens with an invocation or statement about God's
character/action
- Includes a responsive element (leader/congregation
back-and-forth)
- Connects to the day's Scripture or theme
- Uses inclusive, accessible language
- Builds toward a sense of anticipation or
celebration
- Ends with a unison affirmation or invitation

Theological tradition: [Wesleyan/Reformed/etc.]
Tone: Reverent but accessible, invitational
Length: 100-150 words total
Format: Clearly mark Leader and People sections
```

What to Expect

AI will generate a responsive call to worship with leader and congregation parts, connected to your theme and Scripture. The language will be reverent but accessible, suitable for contemporary worship contexts.

Follow-Up Prompts

"Make this call to worship more celebratory and joyful."

This adjusts the emotional tone to match seasons like Easter or Pentecost.

"Rewrite this to use more concrete, sensory language instead of abstract concepts."

You can make the liturgy more vivid and embodied rather than purely intellectual.

"Create a shorter version (50-75 words) for a less formal worship setting."

Sometimes brevity fits your congregation's style better than longer liturgical forms.

How to Use It

1. Read the call to worship out loud to test its flow and rhythm
2. Adjust any language that doesn't match your congregation's vocabulary
3. Ask AI to adjust the emotional tone (more celebratory, more reflective) to match your service
4. Have AI create shorter or longer versions depending on your worship style

Reflection Question

What's the emotional or spiritual state of your congregation as they arrive on Sunday morning? How can this call to worship meet them where they are and gently invite them into worship?

CHAPTER 8
Creating Pastoral Prayers for Sunday Worship

Our Goal
Craft pastoral prayers that intercede for *your* congregation, *your* community, and the world while connecting to the day's theme and Scripture reading.

The Problem
Pastoral prayers need to hold many things at once: congregational joys and concerns, global crises, seasonal themes, and the day's Biblical focus. Praying extemporaneously is an option, but a well-crafted written prayer can offer depth, poetry, and inclusivity that's sometimes hard to achieve in the moment – especially when you're exhausted or distracted.

Why This Matters
The pastoral prayer is one of the most intimate moments in worship. It's when the congregation hears their pastor name their griefs, hopes, fears, and gratitude before God. A thoughtful prayer validates people's experiences, broadens their concern for the world, and models what it sounds like to talk to God. AI can help you draft prayers that are comprehensive and beautiful without spending an hour wordsmithing.

The Prompt
```
I'm writing a pastoral prayer for [season/Sunday:
e.g., Third Sunday in Lent, Ordinary Time] based on
the theme of [theme from your sermon].

Create a pastoral prayer (250-300 words) that
includes:
- An opening address to God
- Thanksgiving for specific blessings (seasonal,
congregational, or Biblical)
- Intercession for the congregation's current needs:
[briefly list: e.g., illness, job loss, grief]
- Intercession for the broader community and world:
[briefly mention: e.g., natural disaster, political
tension, ongoing crisis]
- Connection to the day's Scripture or theme
- A closing that invites trust, hope, or commitment

Theological tradition: [Wesleyan/Reformed/etc.]
Tone: Intimate, honest, hopeful
```

Avoid: Preachy prayers that sound like sermons, vague
generalities, or guilt-inducing language

What to Expect
AI will generate a comprehensive pastoral prayer with clear sections for
thanksgiving, intercession, and closing. The language will be reverent
but personal, and the prayer will balance congregational concerns with
broader world issues.

Follow-Up Prompts
"Add a section specifically praying for children and
families in our congregation."
This helps you tailor the prayer to groups that might otherwise feel
overlooked.
"Make the language more poetic and metaphorical
rather than straightforward prose."
You can adjust the literary style to match your worship tradition and
personal voice.
"Rewrite the intercession section to focus more on
hope and gratitude than on problems and needs."
Sometimes your congregation needs a prayer that lifts their eyes rather
than dwelling on difficulty.

🔒 **Privacy Reminder**: Never name individuals or share specific
confidential details in public pastoral prayers without explicit
permission. Use general categories ("those facing illness," "families
experiencing loss") rather than identifying people.

How to Use It
1. Use AI's prayer as your structural starting point and add specific
   community or congregational details
2. Ask AI to adjust sections that feel too formal or too casual for
   your congregation
3. Have AI expand or condense specific prayer sections to match
   your timing needs

Reflection Question
What unspoken burden or hidden struggle might someone in your
congregation be carrying this Sunday? How can you name it in this
prayer without violating anyone's privacy?

CHAPTER 9
Drafting Prayers for Specific Occasions

Our Goal
Create pastoral prayers for weddings, funerals, baptisms, home blessings, hospital visits, and other occasional services that feel personal and theologically grounded.

The Problem
Every wedding, funeral, and baptism is unique. You want prayers that honor the specific people and circumstances, but writing original prayers for every occasion adds hours to your schedule. Generic prayers from a book feel impersonal.

Why This Matters
Occasional services are some of the most memorable moments in people's lives. A well-crafted prayer acknowledges the significance of the moment and connects their experience to God's presence. AI can help you draft prayers that are both personal and beautiful without starting from scratch.

The Prompt
```
I'm preparing a prayer for [occasion: wedding,
funeral, baptism, home blessing, hospital visit,
etc.] involving [brief, non-identifying description:
e.g., a couple in their 30s, a long-time church
member who died after a long illness, parents
baptizing their first child].

Create a prayer (150-200 words) that:
- Addresses God appropriately for this occasion
- Acknowledges the significance of this moment
- Includes specific intercessions relevant to the
occasion
- Offers hope, blessing, or comfort appropriate to
the context
- Ends with a strong affirmation or benediction

Theological tradition: [Wesleyan/Reformed/etc.]
Tone: [reverent and joyful for weddings/baptisms,
compassionate and hopeful for funerals/hospital
visits]
Avoid: Clichés, overly formal language, or
theological jargon
```

What to Expect
AI will generate a complete prayer appropriate to the occasion with clear structure, theologically sound language, and an emotional tone that matches the event. It will feel somewhat generic until you personalize it.

Follow-Up Prompts
"Make this prayer more personal by referencing [specific detail: e.g., the couple's love of music, the deceased's service to the community]."
Adding specific details transforms a generic prayer into something that could only be prayed for these people.
"Adjust the tone to be more hopeful and celebratory rather than somber."
Different funerals require different emotional registers – a long life well-lived feels different than a tragic, untimely death.
"Add a section that acknowledges grief/difficulty while still affirming God's presence."
You can help the prayer hold both lament and hope in appropriate balance.

🔒 **Privacy Reminder**: When drafting prayers for hospital visits or pastoral care situations, never input identifying details or confidential medical information into AI. Use general descriptions like "someone facing surgery" or "a family in crisis."

🙂 **Your Discernment Matters**: AI can draft beautiful language, but it doesn't know the people you're praying for. Always personalize with specific names, relationships, stories, and details that honor the individuals involved.

How to Use It
1. Use AI's prayer as a structural starting point
2. Add the person's name(s) and specific details about their story
3. Have AI adjust theological language to match the family's faith background

Reflection Question
What do you know about this person, couple, or family that AI could never know? How can you weave that knowledge into the prayer so it becomes a sacred gift tailored specifically to them?

CHAPTER 10
Planning Worship Services (Order, Flow, Themes)

Our Goal
Design cohesive worship services where every element – Scripture, sermon, music, prayers, liturgy – connects thematically and flows smoothly from beginning to end.

The Problem
Worship planning involves coordinating multiple elements: selecting hymns, choosing Scripture readings, writing prayers, planning transitions, and ensuring everything connects to your sermon theme. It's like conducting an orchestra where you're also composing the music. By the time you've planned a coherent service, hours and days have passed.

Why This Matters
A great worship service feels seamless. Each element builds on the last, creating a unified experience that helps people encounter God. When AI helps you plan the overall service structure, you can focus your energy on the pastoral and theological depth rather than the logistics of what goes where.

The Prompt
```
I'm planning worship for [season/Sunday: e.g., Third
Sunday of Advent] based on [Scripture passage] with
the sermon theme of [brief theme description].

Create a worship service outline that includes:
- Order of worship with 8-10 elements (gathering,
hymns, prayers, Scripture, sermon, response, sending)
- Suggested hymn titles or themes for each song
- Brief descriptions of what each element should
accomplish
- Natural transitions between elements
- A flow that builds toward and flows from the sermon
- Balance between participation and listening, joy
and reflection

Worship style: [traditional/contemporary/blended]
Congregation size and context: [description]
Length: Full service outline, about 300-400 words
```

What to Expect
AI will generate a complete worship order with suggested elements, hymn themes, and brief descriptions of each section's purpose. The flow will feel logical and thematically cohesive, though you'll need to select specific hymns and adapt to your congregation's preferences.

Follow-Up Prompts
"Suggest three specific hymn titles for each musical moment based on [hymnal or worship resource you use]."
This helps you move from general hymn themes to actual song selections.
"Add more congregational participation elements – I want people actively involved, not just listening."
You can adjust the balance between liturgy the pastor leads and liturgy the congregation participates in.
"Shorten this service to 60 minutes instead of 75 minutes – what should I cut or condense?"
Sometimes you need to fit worship into a tighter timeframe without losing coherence.

How to Use It
1. Review AI's worship flow for theological coherence and emotional pacing
2. Ask AI to suggest specific hymn titles from your church's hymnal or worship resource
3. Have AI adjust the service length by condensing or expanding specific elements

Reflection Question
Looking at this worship outline, where will people have space to breathe, reflect, and respond to God? Where might the pacing feel rushed or overwhelming? How can you build in moments of silence or contemplation?

CHAPTER 11
Just for Methodists: Ask Wes!

Our Goal
Get quick, reliable answers to United Methodist Church theology, polity, and practice questions using the custom ChatGPT chatbot "Wes" (short for Wesley). You only need a **free** ChatGPT account to use "Wes."

The Problem
United Methodist pastors face constant questions about denominational procedures, Book of Discipline regulations, theological distinctives, and worship practices. You need answers quickly – for a trustee meeting tonight, a confused confirmation student tomorrow, or your own sermon prep right now. Digging through the Book of Discipline takes time you don't have.

Why This Matters
"Wes" is a custom AI chatbot trained specifically on United Methodist resources, theology, and polity. It's like having a denominational expert available 24/7. Whether you're navigating a governance question, exploring Wesleyan theology, or planning a connectional ministry event, Wes can give you informed answers grounded in Methodist tradition.

The Prompt
Access Wes at: **https://chatgpt.com/g/g-D1iJ6slXy-wes**

```
I'm a United Methodist pastor with a question about
[topic: e.g., baptismal theology, trustee
responsibilities, apportionment policies, Wesleyan
view of sanctification].

[Ask your specific question clearly and directly]

Context: [Briefly describe why you're asking – e.g.,
preparing a sermon, answering a church member's
question, navigating a board meeting decision]
```

What to Expect
Wes will provide answers grounded in UMC theology, the *Book of Discipline*, and denominational resources. Responses will cite relevant passages or church documents when applicable. The answers will be more specific to UMC context than general AI chatbots.

Follow-Up Prompts
"Can you cite the specific paragraph in the Book of Discipline that addresses this?"
This helps you get the exact reference you need for official church business or serious theological discussions.
"Explain this in simpler language for a confirmation class."
Wes can adjust theological complexity to match your audience – whether it's youth, new members, or seminary-trained colleagues.
"How does the UMC's position on this differ from other Protestant denominations?"
You can explore denominational distinctives to better understand what makes Methodist theology and practice unique.

⚠ **Verify This**: While Wes is trained on UMC resources, always verify critical information – especially regarding polity, governance, or official church positions – by checking the newest actual *Book of Discipline* or consulting your District Superintendent. AI can occasionally misinterpret complex regulatory language.

✝ **Theological Caution**: Wes reflects mainstream UMC theology, but theology is diverse even within denominations. Use Wes as a starting point for exploration, not as the final word on contested theological questions. Remember, as Methodists, we use the Quadrilateral to help us discern our opinions.

How to Use It
1. Start with clear, specific questions rather than vague theological inquiries
2. When Wes cites *Book of Discipline* paragraphs, verify them in your actual copy
3. Use Wes to explore multiple perspectives on contested issues within the UMC
4. Share Wes with lay leaders who have UMC-specific questions
5. Tell Wes, "Show me how to use the Wesleyan Quadrilateral to address _______ issue."

Reflection Question
What UMC theology, practice, or polity question have you been meaning to research but haven't had time? How might quick access to Wes help you better serve your congregation's questions about Methodist identity and tradition?

SECTION 2

THE LITURGICAL CALENDAR & SPECIAL DAYS

CHAPTER 12
The Liturgical Cycle: Sermon Planning for the Year

Our Goal
Create a year-long preaching plan that follows the liturgical calendar, balances Scripture coverage, and addresses your congregation's spiritual needs across seasons.

The Problem
Planning sermons week-by-week feels reactive and exhausting. You're always scrambling for next Sunday's topic, missing opportunities to build thematic momentum, and sometimes avoiding difficult texts because they feel inconvenient. A year-long plan provides structure, ensures Biblical breadth, and lets you prepare well in advance – but creating that plan from scratch is overwhelming.

Why This Matters
Following the liturgical cycle roots your preaching in the church's ancient wisdom about spiritual formation. Advent prepares hearts for Christ's coming. Lent leads toward Easter. Ordinary Time allows sustained exploration of discipleship. When AI helps you map out a year of preaching that follows this rhythm, you gain both freedom (from weekly panic) and faithfulness (to the full scope of the Biblical narrative and Christian year).

The Prompt
```
I'm a [denomination] pastor planning a year of
preaching that follows the liturgical calendar from
[start date: e.g., First Sunday of Advent] through
[end date: e.g., Christ the King Sunday].

Create a preaching plan that includes:
- Sermon series for each liturgical season (Advent,
Christmas, Epiphany, Lent, Easter, Pentecost,
Ordinary Time)
- Suggested themes and Scripture passages for each
season
- Balance between Old Testament, Gospels, Epistles,
and Psalms
- Integration of major holy days (Ash Wednesday, Palm
Sunday, Maundy Thursday, Good Friday, Easter,
Pentecost, All Saints, etc.)
```

```
- 2-3 topical series during Ordinary Time that
address congregational needs: [list any specific
topics you know you need to address]

Congregation context: [size, demographics, spiritual
maturity, current challenges]
Theological tradition: [Wesleyan/Reformed/etc.]
Lectionary use: [following Revised Common Lectionary,
selective use, or independent planning]

Length: Year-long outline with brief descriptions for
each series/season
```

What to Expect
AI will generate a comprehensive year-long preaching calendar
organized by liturgical seasons, with sermon series suggestions, Scripture
coverage, and topical focuses during Ordinary Time. The plan will
balance liturgical rhythm with congregational needs.

Follow-Up Prompts
```
"Expand the Lent series into week-by-week sermon
titles and Scripture passages."
```
Once you have the big picture, you can drill down into specific seasons
with more detail.
```
"Suggest where in this plan I could address [specific
congregational issue: e.g., financial generosity,
racial reconciliation, mental health awareness]."
```
You can strategically place needed topics within the liturgical framework
rather than treating them as interruptions.
```
"Adjust this plan to include monthly Communion
services and quarterly special Sundays (UMC
apportionments, mission emphasis, etc.)."
```
Denominational and sacramental rhythms can be woven into the
preaching calendar seamlessly.

📖 **Scripture Check**: Always verify your AI is looking at the correct
liturgical year (A, B, or C). If you're planning a year in advance, you'll
need to explain that in your prompts.

✅ **Pro tip:** Google "Revised Common Lectionary for all years" to find a
downloadable PDF of the lectionary year you want. Upload that file to
your AI to ensure you're both looking at the same list of verses and
themes.

How to Use It
1. Review AI's year-long plan for theological balance and congregational fit
2. Use AI to drill down into specific seasons by requesting week-by-week details
3. Have AI adjust the plan when unexpected events require topical preaching

Reflection Question

Looking at this year-long plan, which season or series excites you most? Which one feels most daunting? How can you prepare emotionally and spiritually for the hard parts, not just the content?

CHAPTER 13
Advent: Hope, Peace, Joy, and Love

Our Goal
Create a cohesive four-week Advent sermon series and worship plan that explores the traditional themes of hope, peace, joy, and love while preparing hearts for Christmas.

The Problem
Advent is liturgically rich but logistically demanding. You need four sermons that connect thematically, worship elements that reflect each week's focus, and a sense of progression that builds anticipation for Christmas. Meanwhile, you're also managing increased holiday demands – Christmas programs, extra services, year-end giving campaigns, and family expectations. Planning it all feels overwhelming.

Why This Matters
Advent prepares hearts for Christmas. A well-planned series helps your congregation move beyond cultural Christmas chaos into spiritual anticipation. When AI handles the planning framework, you can focus on adding the pastoral depth and local context that make Advent meaningful in your specific community.

The Prompt
```
I'm a [denomination] pastor planning a four-week
Advent sermon series. Our congregation is
[description].

Create an Advent series plan that includes:

**Week 1 (Hope):**
- Sermon title and Scripture passage
- 3 main sermon points
- Call to worship
- Suggested hymn/song

**Week 2 (Peace):**
- Sermon title and Scripture passage
- 3 main sermon points
- Call to worship
- Suggested hymn/song

**Week 3 (Joy):**
- Sermon title and Scripture passage
```

```
- 3 main sermon points
- Call to worship
- Suggested hymn/song

**Week 4 (Love):**
- Sermon title and Scripture passage
- 3 main sermon points
- Call to worship
- Suggested hymn/song


Theological perspective: [Wesleyan/Reformed/etc.]
Tone: Anticipatory, reflective, hopeful
Include: Balance between traditional Advent themes
and contemporary application
```

What to Expect

AI will generate a complete four-week framework with thematically connected sermons, Scripture suggestions, worship elements, and hymn ideas. The series will feel cohesive but may need adjustments to match your congregation's musical preferences and spiritual needs.

Follow-Up Prompts

```
"Adjust Week 3 (Joy) to address grief during the
holidays for those who are struggling."
```
This recognizes that not everyone experiences joy during this season – AI can help you hold space for both celebration and sorrow.
```
"Suggest contemporary worship songs instead of
traditional hymns."
```
You can quickly adapt the musical style to match your congregation's preferences.
```
"Add a children's message idea for each week that
connects to the Advent theme."
```
This expands your framework to serve multiple age groups without starting from scratch.

📖 **Scripture Check**: Verify that suggested Advent passages actually support the weekly theme. Some traditional "joy" texts may be more complex than they appear at first glance.

How to Use It

1. Review AI's Scripture choices for theological soundness
2. Ask AI to suggest contemporary songs if traditional hymns don't fit your congregation

3. Have AI create children's messages for each week that connect to the Advent themes

Reflection Question
Which week's theme will be most challenging for your congregation this year based on what they're experiencing? How might you adjust that sermon to meet them where they are?

CHAPTER 14
Christmas Eve: Candlelight Service Planning

Our Goal
Design a Christmas Eve candlelight service that balances tradition, accessibility for visitors, and meaningful worship for your congregation.

The Problem
Christmas Eve is one of the most attended services of the year. Your sanctuary will be full of regular members, visiting family, people who only come once or twice a year, and neighbors curious about church. You need a service that honors tradition without alienating newcomers, that feels sacred without being stuffy, and that tells the Christmas story in a way that resonates with everyone.

Why This Matters
Christmas Eve is often people's first or only encounter with your church. The service sets the tone for how they experience Christian worship and whether they'll consider returning. A well-planned Christmas Eve service creates space for both reverent worship and warm welcome, making room for God to meet people wherever they are in their faith journey.

The Prompt
I'm planning a Christmas Eve candlelight service for a [size] congregation with a mix of regular members and visitors.

Create a worship service outline (60-75 minutes) that includes:
- Opening that warmly welcomes both members and first-time visitors
- Scripture readings from the Christmas narrative (Luke 2, other passages)
- 3-4 traditional Christmas hymns/carols
- A brief, accessible sermon (10-12 minutes) on [theme: e.g., incarnation, God-with-us, hope in darkness]
- Candlelight ritual with explanation for those unfamiliar
- Closing carol and benediction

Worship style: [traditional/contemporary/blended]
Tone: Reverent, warm, accessible to spiritual seekers

```
Include: Balance between familiar tradition and
welcoming explanation
```

What to Expect
AI will generate a complete service outline with suggested Scripture readings, carol selections, a brief sermon structure, and clear instructions for the candlelight portion. The flow will feel traditional yet accessible, suitable for both longtime members and first-time visitors.

Follow-Up Prompts
```
"Add a children's moment or interactive element that
engages families with young kids."
```
This helps you include families without making the service feel juvenile to adults.
```
"Suggest specific carol verses to sing rather than
all verses – I need to keep the service under 70
minutes."
```
You can tighten the timing without losing the beloved music.
```
"Write a brief welcome script (100 words) that makes
first-time visitors feel genuinely included."
```
The opening welcome sets the tone for whether visitors feel like outsiders or honored guests.

How to Use It
1. Use AI's service outline as your starting framework
2. Ask AI to write a visitor-focused welcome script for your specific context
3. Have AI create visitor-focused social media text and graphics based on your final outline

Reflection Question
Who might walk through your doors on Christmas Eve carrying grief, loneliness, or doubt? How can your service create space for both their pain and God's comfort without pretending everything is merry and bright?

CHAPTER 15
Christmas Day: Incarnation-Focused Worship

Our Goal
Create a Christmas Day worship service that goes deeper into the theology of incarnation for those who attend beyond Christmas Eve.

The Problem
Christmas Day worship serves a different audience than Christmas Eve. These are your committed members, people staying in town, and those who want more than candlelight sentimentality. They've already heard the familiar Christmas story – now they want to dig into what it actually means that God became flesh. But after the emotional high of Christmas Eve, you're exhausted and still need to lead meaningful worship.

Why This Matters
Christmas Day is an opportunity to move from "baby Jesus in a manger" to the radical theological claim of incarnation: that God entered human existence, took on flesh, and dwelt among us. This service can explore the deeper "why" questions that Christmas Eve doesn't always address. When AI helps you plan this service, you can create worship that honors both the joy of the season *and* the profound mystery of God-with-us.

The Prompt
```
I'm planning a Christmas Day worship service for
church members who want theological depth beyond the
traditional Christmas Eve story.

Create a worship service outline (60 minutes) that
includes:
- Opening that acknowledges both the joy of Christmas
and the deeper mystery of incarnation
- Scripture readings that explore incarnation
theology (John 1:1-14, Philippians 2:5-11, or other
passages)
- 3-4 hymns that focus on incarnation rather than
just nativity sentimentality
- Sermon structure (15-20 minutes) on [theme: e.g.,
what it means that God became flesh, why incarnation
matters, Emmanuel – God with us in suffering and joy]
- Prayer of thanksgiving and commitment
- Sending that connects Christmas joy to ongoing
discipleship
```

```
Theological tradition: [Wesleyan/Reformed/etc.]
Tone: Joyful but reflective, theologically grounded
```

What to Expect
AI will generate a worship outline with deeper theological focus, Scripture that explores incarnation beyond the nativity story, and sermon structure that moves from "Christmas happened" to "here's why incarnation changes everything." The tone will balance celebration with theological reflection.

Follow-Up Prompts
```
"Suggest specific hymns from [your hymnal] that focus
on incarnation theology."
```
This helps you find the right music in the resources you actually have available.
```
"Add a responsive reading or litany that names the
ways God-with-us matters in our daily lives."
```
You can make the incarnation feel personally relevant rather than just doctrinally important.
```
"Include a brief Communion liturgy that connects the
incarnation to the Eucharist."
```
Christmas Day is a meaningful time to celebrate the sacrament with those who show up for deeper worship.

How to Use It
1. Use AI's outline but ask it to suggest hymns specifically from your hymnal
2. Ask AI to create a responsive reading about incarnation for congregational participation
3. Have AI develop a brief Communion liturgy that connects incarnation to Eucharist

Reflection Question
For the people who show up on Christmas Day, what are they hungry for spiritually that Christmas Eve didn't fully satisfy? How can you feed that hunger?

CHAPTER 16
Epiphany: Light of the World

Our Goal

Plan an Epiphany worship service that celebrates Jesus as the light revealed to all nations and explores what it means to live as light-bearers in the world.

The Problem

Epiphany often gets lost between Christmas and the return to Ordinary Time. It's not a major cultural holiday, so it doesn't command attention like Christmas or Easter. But Epiphany carries crucial theology: the light of Christ isn't just for insiders – it's revealed to the whole world, symbolized by the Magi's journey. Creating worship that makes Epiphany feel significant requires intentional planning.

Why This Matters

Epiphany is about mission, revelation, and God's love for all people. It's when we celebrate that Jesus came not just for the faithful Jews waiting for Messiah, but for Gentile astrologers from the East – and by extension, for *everyone*. Epiphany worship reminds us that we're called to be light-bearers, carrying the revelation of Christ into the world. When AI helps you plan this service, you can create worship that recovers Epiphany's theological richness.

The Prompt

```
I'm planning an Epiphany worship service for [date]
focusing on the theme of Christ as light revealed to
the world.

Create a worship service outline that includes:
- Opening that introduces Epiphany's significance
(many don't know what it means)
- Scripture readings from the Epiphany narrative
(Matthew 2:1-12, Isaiah 60:1-6, or Ephesians 3:1-12)
- 3-4 hymns focused on light, revelation, mission, or
the Magi's journey
- Sermon structure (15-20 minutes) on [theme: e.g.,
God's love revealed to all nations, we are called to
be light-bearers, the journey of the Magi as
spiritual metaphor]
- A symbolic element (lighting candles, star imagery,
gift-bringing ritual)
```

```
- Sending that commissions worshipers to carry Jesus'
light into the world

Worship style: [traditional/contemporary/blended]
Theological tradition: [Wesleyan/Reformed/etc.]
Tone: Celebratory, missional, globally minded
```

What to Expect
AI will generate a complete service outline with Epiphany-focused
Scripture, hymn suggestions emphasizing light and revelation, sermon
structure, and ideas for symbolic elements. The service will help people
understand why Epiphany matters and what it calls them to do.

Follow-Up Prompts
```
"Add a global or multicultural element that reflects
the 'all nations' theme of Epiphany."
```
This helps you embody the theological claim that Jesus' light is for
everyone, not just one culture.
```
"Suggest a children's activity or object lesson using
stars or light that teaches the Epiphany story."
```
You can make the Magi's journey accessible to kids while preserving its
theological depth for adults.
```
"Create a commissioning prayer or litany that sends
people out as light-bearers."
```
The service can end with a clear call to mission rather than just
celebrating a historical event.

How to Use It
1. Use AI to add multicultural elements that reflect the "all nations"
   theme
2. Ask AI to create a children's activity using stars or light
3. Have AI write a commissioning prayer or benediction that sends
   people as light-bearers

Reflection Question
Who are the "outsiders" or "unexpected seekers" in your community –
the people your congregation might not expect to be drawn to Christ's
light? How can this Epiphany service remind your people that God's
love reaches beyond their comfort zone?

CHAPTER 17
Martin Luther King Jr. Day: Ecumenical Services and Prayers

Our Goal
Create worship elements for Martin Luther King Jr. Day observances that honor Dr. King's legacy, address racial justice, and connect to Biblical themes of beloved community.

The Problem
Many churches participate in ecumenical MLK Day services or hold their own observances. You want to honor Dr. King's prophetic witness and address ongoing racial injustice, but you're navigating sensitive territory – varying political perspectives in your congregation, the risk of tokenizing Black voices, and the challenge of moving beyond nice words to actual commitment. Planning worship that is both faithful and pastorally wise requires careful thought.

Why This Matters
Dr. King's vision of the beloved community is deeply Biblical. Honoring his legacy isn't about politics – it's about the Gospel's call to justice, reconciliation, and the dignity of all people made in God's image. MLK Day services can help congregations lament ongoing racial injustice, celebrate progress, and recommit to the work of dismantling racism.

The Prompt
```
I'm planning worship elements for a Martin Luther
King Jr. Day observance [specify: ecumenical
community service, church worship service, or prayer
vigil].

Create worship materials that include:
- A call to worship that connects Dr. King's vision
to Biblical themes (justice, beloved community,
reconciliation)
- Scripture readings that ground racial justice in
the Gospel (options: Micah 6:8, Amos 5:24, Galatians
3:28, Luke 4:18-19)
- A prayer of confession that names complicity in
racism and calls for repentance
- A prayer of commitment to work for racial justice
and reconciliation
- Suggested hymns or songs (both traditional justice
hymns and contemporary worship)
```

```
- [Optional: Create a brief sermon outline on beloved
community, prophetic witness, or ongoing racial
justice work]

Context: [Describe your congregation or community -
racial diversity, location, ecumenical partnerships]
Theological tradition: Wesleyan
Tone: Prophetic but pastoral, honest about ongoing
injustice, hopeful without being naïve
```

What to Expect

AI will generate worship elements grounded in Biblical justice themes, prayers that acknowledge both lament and hope, and suggestions for music that honors the civil rights movement while being accessible to diverse congregations. The materials will balance honoring Dr. King's legacy with calling people to ongoing action.

Follow-Up Prompts

```
"Adjust the prayer of confession to be more specific
about systems of injustice (housing, education,
policing, healthcare)."
```
Moving from vague acknowledgment to concrete naming of injustice can deepen congregational awareness.
```
"Add a responsive reading that includes quotes from
Dr. King alongside Scripture."
```
Weaving Dr. King's words with Biblical text honors his prophetic voice as part of the church's ongoing witness.
```
"Suggest ways to connect this worship service to
concrete action steps our congregation can take."
```
Worship should lead to discipleship – AI can help you identify tangible commitments beyond the service itself.

Your Discernment Matters: Worship services focused on racial justice require deep pastoral sensitivity. void using Black voices as symbolic gestures without genuine partnership, oversimplifying complex issues, or treating this as a one-day-a-year concern. Consider partnering with diverse church leaders in your community and ensuring that this observance connects to ongoing antiracism work in your congregation.

How to Use It

1. Customize AI's outline with actual stories from your community's racial justice work

2. Ask AI to create a litany weaving Dr. King's words with Scripture
3. Have AI suggest concrete action steps your congregation can take to address racism in your community

Reflection Question

How is your congregation currently engaged (or *not* engaged) in addressing racism beyond this annual observance? What would it look like to make Dr. King's vision of beloved community a year-round commitment rather than a one-day commemoration?

CHAPTER 18
Ash Wednesday: Confession and Lament

Our Goal
Create an Ash Wednesday service that invites honest confession, embraces the discipline of Lent, and marks the transition from ordinary life into this sacred season of preparation.

The Problem
Ash Wednesday is countercultural. In a society that avoids talking about mortality and sin, you're asking people to hear "remember that you are dust, and to dust you shall return" while receiving ashes on their foreheads. The service needs to be solemn without being guilt-inducing, honest about human brokenness without being despairing, and invitational rather than mandatory. Creating worship that holds all of this together is emotionally and liturgically demanding.

Why This Matters
Ash Wednesday begins the Lenten journey. It names the reality of our mortality, invites confession of sin, and calls us into spiritual disciplines that prepare us for Easter. This service creates space for people to acknowledge what they usually avoid: that we are finite, broken, and in need of God's grace. When AI helps you plan Ash Wednesday worship, you can create a service that is both appropriately solemn and pastorally tender.

The Prompt
```
I'm planning an Ash Wednesday service that marks the
beginning of Lent with confession, imposition of
ashes, and a call to spiritual disciplines.

Create a worship service outline (45-60 minutes) that
includes:
- Opening that explains Ash Wednesday's purpose and
invites people into Lent
- Scripture readings focused on repentance,
mortality, and God's mercy (Joel 2:1-2, 12-17; Psalm
51; Matthew 6:1-6, 16-21; or 2 Corinthians 5:20b-
6:10)
- 2-3 somber, reflective hymns (no "alleluias" during
Lent)
```

```
- Brief sermon (10-12 minutes) on [theme: e.g.,
embracing our mortality, the gift of repentance,
beginning the Lenten journey]
- Prayer of confession (corporate)
- Imposition of ashes with clear instructions for
those unfamiliar with the practice
- Introduction of Lenten disciplines (prayer,
fasting, service)
- Quiet closing without triumphant sending

Theological tradition: [Wesleyan/Reformed/etc.]
Tone: Solemn, honest, tender
Avoid: Guilt-inducing language, despair without hope,
making ashes feel like punishment
```

What to Expect

AI will generate a complete Ash Wednesday service with appropriate Scripture, somber hymn suggestions, sermon structure, confession liturgy, and guidance for the imposition of ashes. The tone will be serious but not punishing, inviting people into spiritual honesty.

Follow-Up Prompts

```
"Write a brief explanation (100 words) of the
imposition of ashes for people who've never
experienced this practice."
```
Many Protestants are unfamiliar with receiving ashes – clear explanation reduces anxiety and increases participation.
```
"Create a longer, more poetic prayer of confession
that names specific sins and brokenness."
```
Corporate confession can move beyond vague generalities to name real human failings.
```
"Add a brief Communion liturgy – we observe Ash
Wednesday as a day of fasting and Eucharist."
```
Some traditions pair the ashes with the Lord's Supper as a sign of God's grace even in our mortality.

 Your Discernment Matters: Ash Wednesday addresses mortality and sin directly. Be pastorally sensitive to those carrying fresh grief, struggling with mental health, or dealing with shame. The goal is honest confession that leads to grace, not condemnation that leads to despair.

 Pro tip: Mix ashes with oil, *not* water. Ashes + water = lye. Lye is a caustic, alkaline solution that burn skin. That's not the mark you want to leave on people's foreheads or hands!

How to Use It
1.   Ask AI to write a brief explanation of imposition of ashes for first-time participants
2.   Use AI to create a longer, more poetic corporate confession
3.   Have AI add a Communion liturgy if your tradition pairs ashes with Eucharist

Reflection Question
What are you personally hoping to confess or lay down during this Lenten season? How can your own spiritual honesty shape the way you lead this service for others?

CHAPTER 19
Lent: 40-Days of Reflections

Our Goal
Create a 40-day Lenten devotional series that helps your congregation journey through Lent with spiritual discipline and deepening faith.

The Problem
Lent is a season of preparation, but most people don't know how to observe it beyond "giving something up." You want to offer your congregation a richer spiritual experience – daily reflections, Scripture readings, prayers, or practices that guide them through the 40 days. But writing 40 days of devotional content from scratch feels impossible.

Why This Matters
A structured Lenten devotional gives your congregation a shared spiritual practice. When everyone is reading the same Scripture and reflecting on the same themes, it builds communal depth. Daily devotionals also extend your pastoral care beyond Sunday – you're walking with people through their week. When AI helps you create this content, you can offer rich spiritual formation without burning out.

The Prompt
```
I'm creating a 40-day Lenten devotional for my
congregation from Ash Wednesday through Easter.

For each day, create a devotional entry (150-200
words) that includes:
- A brief Scripture passage (vary between Gospels,
Psalms, prophets, and Epistles)
- A short reflection that connects the Scripture to
the Lenten themes of repentance, sacrifice,
preparation, and hope
- A closing prayer or spiritual practice prompt
- [Optional: A question for personal reflection or
journaling]

Theological tradition: Wesleyan
Congregation context: [describe your congregation -
age, spiritual maturity, life circumstances]
Tone: Reflective, honest, hopeful
Themes to emphasize: [e.g., God's mercy, the cost of
discipleship, dying and rising with Christ,
preparation for Easter]
```

Start with Day 1 (Ash Wednesday) and create entries
for the first week. I'll request subsequent weeks as
we progress.

What to Expect

AI will generate devotional entries with Scripture, reflection, and prayer
for each day. The reflections will be theologically sound and pastorally
warm, though you'll want to personalize them with local stories or
congregational specifics. You can request devotionals in batches (week
by week) rather than all 40 at once. Doing so gives you the opportunity
to update your prompts so your AI generates content that sounds more
like you and meets your specific, contextual needs.

Follow-Up Prompts

"Create Week 2 devotionals following the same format,
focusing on the theme of [wilderness, temptation,
trust, etc.]."

You can build thematic progression across the weeks while maintaining
consistency in format.

"Add a specific spiritual practice suggestion for
each day (silence, fasting, acts of service,
gratitude journaling, etc.)."

This moves people beyond just reading to actually practicing Lenten
disciplines.

"Adjust the reading level to be appropriate for
teenagers – I want youth to use this too."

You can create versions for different age groups or spiritual maturity
levels.

How to Use It

1. Request devotionals from AI in weekly batches rather than all 40
   at once
2. Ask AI to add specific spiritual practices for each day beyond
   just reading
3. Have AI create versions for different age groups or spiritual
   maturity levels

Reflection Question

What spiritual discipline or practice do you personally need during this
Lenten season? How can you integrate your own Lenten journey into the
devotionals you offer your congregation?

CHAPTER 20
Palm Sunday: Celebration

Our Goal
Plan a Palm Sunday worship service that celebrates Jesus' triumphal entry while foreshadowing the suffering and death that follows, holding both joy and sorrow in tension.

The Problem
Palm Sunday is liturgically complex. It's a day of celebration – waving palms, shouting "Hosanna," welcoming the King. But it's also the doorway into Holy Week, and crowds shouting praises on Sunday will shout "Crucify him!" by Friday. You need worship that honors the joy without ignoring the tragedy ahead.

Why This Matters
Palm Sunday teaches us that faithful discipleship means following Jesus even when the parade turns into a crucifixion march. Your congregation needs to hear both the "Hosannas" and the hard truth that following Jesus leads to a cross. When AI helps you plan this service, you can create worship that holds the tension without collapsing into either uncomplicated joy or premature grief.

The Prompt
```
I'm planning a Palm Sunday worship service that
celebrates Jesus' triumphal entry while foreshadowing
the passion and death ahead.

Create a worship service outline (60-75 minutes) that
includes:
- A processional with palm branches and
congregational participation (instructions for those
unfamiliar)
- Scripture reading of the triumphal entry (Matthew
21:1-11, Mark 11:1-11, Luke 19:28-40, or John 12:12-
16)
- 2-3 celebratory hymns ("All Glory, Laud and Honor,"
"Hosanna, Loud Hosanna," or contemporary equivalents)
- Brief sermon (12-15 minutes) that moves from
celebration to the sobering reality of Holy Week
ahead
- [Optional: Dramatic reading or passion narrative
excerpt that shifts the tone]
- Prayers that hold both joy and lament
```

```
- Invitation to Holy Week services (Maundy Thursday,
Good Friday, Easter Vigil)
- Sending that acknowledges the journey isn't over -
we're heading toward the cross


Worship style: [traditional/contemporary/blended]
Theological tradition: Wesleyan
Tone: Celebratory at first, shifting to reflective
and somber
```

What to Expect
AI will generate a service outline with palm processional, celebratory opening, Scripture readings, hymn suggestions, and sermon structure transitioning from triumph to the cross.

Follow-Up Prompts
```
"Add a children's moment that explains why we wave
palms and what happens next in the story."
```
Kids can participate fully in the celebration while being prepared for the hard parts of the story ahead.
```
"Suggest a dramatic reading or responsive Psalm that
shifts the tone from celebration to lament."
```
A liturgical turning point can help the congregation emotionally transition from "Hosanna" to Holy Week.
```
"Include brief descriptions of Maundy Thursday and
Good Friday services so people understand the full
Holy Week journey."
```
Many people only come to Palm Sunday and Easter – you need to invite them into the whole story.

How to Use It
1. Ask AI to create a children's moment explaining palms and what happens next
2. Use AI to suggest a dramatic reading that shifts tone from celebration to Holy Week
3. Have AI write brief descriptions of Maundy Thursday and Good Friday for your bulletin

Reflection Question
How can you help your congregation understand that the same community included both crowds shouting "Hosanna!" as well as crowds shouting "Crucify Him!" – and that your community is that same crowd? Some embrace Jesus and His message while others reject Him.

CHAPTER 21
Maundy Thursday: Last Supper and Foot-Washing Service

Our Goal
Create a Maundy Thursday service that commemorates the Last Supper, explores Jesus' command to love and serve, and incorporates foot-washing or other acts of humble service.

The Problem
Maundy Thursday is intimate and vulnerable. You're often asking people to participate in foot-washing – a deeply physical, potentially awkward act of service – or to witness Communion in a new light as Jesus' final meal with His friends before betrayal and death. The service needs to be reverent without being stiff, participatory without being forced, and emotionally honest about the grief and love intertwined in this night.

Why This Matters
Maundy Thursday takes us into the upper room. We remember Jesus washing His disciples' feet, instituting the Lord's Supper, and giving the new commandment to love one another (John 13:34). This service invites us into servanthood, Eucharistic mystery, and the painful reality that love and betrayal can sit at the same table. When AI helps you plan this worship, you can create a service that is both ancient and contemporary, helping people experience the weight of this holy night.

The Prompt
```
I'm planning a Maundy Thursday service that
commemorates the Last Supper and includes foot-
washing or another act of humble service.

Create a worship service outline (60-75 minutes) that
includes:
- Opening that sets the scene in the upper room
- Scripture readings (John 13:1-17, 31b-35; Exodus
12:1-14; 1 Corinthians 11:23-26; or other Last Supper
texts)
- 2-3 reflective hymns focused on servanthood,
communion, and Jesus' love
- Brief sermon (10-12 minutes) on [theme: e.g., the
new commandment, servant leadership, love at the
table with betrayal]
```

```
- Foot-washing ritual with clear instructions and
pastoral sensitivity with an option for those
uncomfortable participating
- Communion liturgy that emphasizes this as Jesus'
final meal with His friends
- Stripping of the altar or other symbolic act
marking the transition to Good Friday
- Silent or somber departure (no benediction - the
service is unfinished until Easter)

Worship style: [traditional/contemporary/blended]
Theological tradition: Wesleyan
Tone: Intimate, vulnerable, solemn
```

What to Expect

AI will generate a complete service outline with Scripture, hymn suggestions, sermon structure, detailed instructions for foot-washing, Communion liturgy, and guidance for the altar stripping or symbolic transition to Good Friday. The service will feel reverent and participatory.

Follow-Up Prompts

```
"Provide alternative service ideas for congregations
uncomfortable with foot-washing (hand-washing,
serving one another, other acts of humble service)."
```
Not every congregation is ready for foot-washing – AI can suggest meaningful alternatives that preserve the servanthood theme.
```
"Write a brief script explaining the foot-washing
ritual so participants know what to expect and how to
participate."
```
Clear instruction reduces anxiety and helps people enter the ritual with spiritual openness rather than self-consciousness.
```
"Add a section that acknowledges Judas' presence at
the table - the reality of betrayal alongside love."
```
Maundy Thursday holds both deep love and deep pain – naming that tension can make the service more honest.

Your Discernment Matters: Foot-washing is physically and emotionally vulnerable. Provide clear opt-out options for those uncomfortable participating. Be sensitive to people with physical disabilities, trauma histories, or cultural backgrounds where foot-washing carries different meanings. Never force participation.

How to Use It

1. Ask AI for alternative service ideas if your congregation is uncomfortable with foot-washing
2. Use AI to write a script explaining the foot-washing ritual to reduce participant anxiety
3. Have AI add a section acknowledging Judas' presence – the reality of betrayal alongside love

Reflection Question

Who in your life has served you humbly, the way Jesus served His disciples? How does that memory shape your understanding of what Jesus is calling you to do for others?

CHAPTER 22
Good Friday: Tenebrae or Seven Last Words Service

Our Goal
Create a Good Friday service that invites people into the darkness and grief of the crucifixion through Tenebrae (service of shadows) or meditation on Jesus' seven last words from the cross.

The Problem
Good Friday is the hardest day in the Christian year. You're asking people to sit with Jesus' suffering and death, to resist the urge to jump ahead to Easter, and to let themselves feel the weight of the cross. Many congregations avoid Good Friday services altogether because they're uncomfortable with prolonged grief and lament. Creating worship that is appropriately somber without being exploitative or guilt-inducing requires deep pastoral sensitivity.

Why This Matters
Good Friday teaches us that resurrection comes through death, not around it. We can't skip the cross and get to Easter. This service invites us to stay with Jesus in His suffering, to witness the cost of love, and to sit in the darkness before the light comes. When AI helps you plan Good Friday worship, you can create a service that honors the gravity of this day while trusting that Easter / Resurrection Sunday is coming.

The Prompt
```
I'm planning a Good Friday service using [choose one:
Tenebrae format with progressive darkness, meditation
on the Seven Last Words, or passion narrative
reading].

Create a worship service outline (60-90 minutes) that
includes:

**For Tenebrae:**
- 7 Scripture readings tracing Jesus' path to the
cross
- 7 candles or lights that are extinguished one by
one
- Brief reflections or silence after each reading
- Concluding with near-total darkness and departure
in silence
```

```
**For Seven Last Words:**
- Scripture passages for each of Jesus' seven
statements from the cross
- Brief meditations (3-5 minutes each) on the meaning
of each word
- Periods of silence or musical meditation between
each word
- Closing in darkness and silence

Include: Appropriate hymns or choral music, no
"alleluias," prayers that embrace lament, departure
without benediction (the service is unfinished)

Worship style: [traditional/contemporary/blended]
Theological tradition: Wesleyan
Tone: Somber, mournful, honest about suffering
Avoid: Guilt manipulation, graphic violence, despair
without the promise of resurrection
```

What to Expect

AI will generate a complete service outline with Scripture readings, meditation structures, hymn suggestions, and clear instructions for the Tenebrae candle progression or Seven Last Words format. The service will feel reverent, sorrowful, and spiritually honest.

Follow-Up Prompts

```
"Write brief meditations (200 words each) for each of
the Seven Last Words from the cross."
```

This gives you complete sermon content for each of Jesus' final statements, ready to adapt to your voice.

```
"Suggest choral or instrumental music that works
between readings without breaking the somber tone."
```

Music can carry emotion when words feel inadequate, but it needs to match the gravity of the moment.

```
"Add a corporate prayer of lament that acknowledges
both the historical crucifixion and ongoing suffering
in our world."
```

Good Friday connects Jesus' death to present-day injustice, violence, and grief.

🌑 **Your Discernment Matters**: Good Friday worship can be emotionally intense. Be pastorally present for those carrying fresh grief, trauma, or depression. Avoid graphic descriptions of crucifixion violence that exploit suffering rather than honor it. Trust that the darkness of

Good Friday is held by the promise of Easter, even if we don't name it yet.

How to Use It
1. Use AI to write brief meditations (200 words each) for each of the Seven Last Words
2. Have AI suggest choral or instrumental music between readings
3. Ask AI to create a corporate prayer of lament connecting Jesus' death to present suffering

Reflection Question
Where in your life or in the world are you experiencing Good Friday darkness right now? How does staying present to Jesus' suffering help you stay present to real suffering instead of rushing past it toward easy answers?

CHAPTER 23
Easter Sunday: Resurrection Celebration

Our Goal
Plan an Easter Sunday worship service that proclaims the resurrection with joy, triumph, and theological depth while welcoming visitors.

The Problem
Easter is the highest attendance Sunday of the year. Your sanctuary will be full of regular members, visiting family, CEOs (Christmas and Easter Only), and spiritual seekers. You need worship that celebrates Jesus' victory over death without feeling performative or shallow, and that moves from "Jesus is alive" to "so what does that mean for us?"

Why This Matters
Easter is the foundation of Christian faith. "If Christ has not been raised, then our proclamation is in vain and your faith is in vain" (1 Corinthians 15:14, NRSV Up. Ed.). This service proclaims the most audacious claim of the Gospel: death does not have the final word. When AI helps you plan Easter worship, you can create a service that is both joyfully celebratory and theologically grounded.

The Prompt
```
I'm planning an Easter Sunday worship service for a
sanctuary with regular members and many visitors.

Create a worship service outline (75-90 minutes) that
includes:
- Festive opening that immediately announces "Jesus
is risen!"
- Scripture readings from the Easter narrative
(options: Matthew 28, Mark 16, Luke 24, John 20, Acts
10:34-43, 1 Corinthians 15:1-11)
- 4-5 triumphant Easter hymns (traditional and
contemporary options)
- Sermon (15-20 minutes) on [theme: e.g.,
resurrection as God's victory over death, what it
means to live as resurrection people, the empty tomb
changes everything]
- Baptism or reaffirmation of baptismal vows
(optional but traditional for Easter)
- Prayers of thanksgiving and commissioning
- Joyful sending with "alleluias" restored after
Lenten silence
```

```
Worship style: [traditional/contemporary/blended]
Theological tradition: Wesleyan
Tone: Triumphant, joyful, celebratory but not shallow
Include: Welcome for visitors, invitation to ongoing
discipleship
```

What to Expect

AI will generate a complete Easter service with festive opening, Scripture suggestions, hymn recommendations, sermon structure, and liturgical elements. The service will balance celebration with theological substance and include clear welcome for visitors.

Follow-Up Prompts

```
"Add a children's moment that explains the empty tomb
in age-appropriate language with wonder and joy."
```
Easter Sunday often has many children present – an engaging children's moment can capture everyone's attention.
```
"Write a welcome script (150 words) that makes first-
time visitors feel genuinely valued, not just
tolerated."
```
How you welcome people on Easter Sunday may determine whether they return.
```
"Suggest contemporary worship songs that match the
energy of traditional Easter hymns like 'Christ the
Lord Is Risen Today.'"
```
Blended worship requires finding contemporary equivalents that carry the same theological weight and emotional power.

How to Use It

1. Use AI to create a children's moment about the empty tomb with wonder and joy
2. Have AI write a visitor-focused welcome script that makes newcomers feel valued
3. Ask AI to suggest contemporary songs matching the energy of traditional Easter hymns

Reflection Question

How can this service both celebrate resurrection and gently invite them into deeper faith? What's their next step with your congregation?

CHAPTER 24
Pentecost: Holy Spirit and Mission

Our Goal
Create a Pentecost worship service that celebrates the gift of the Holy Spirit, explores the birth of the church, and commissions the congregation for mission in the world.

The Problem
Pentecost often gets overlooked. It's not a cultural holiday like Christmas or Easter, so it doesn't draw crowds. But Pentecost is theologically crucial – it's the outpouring of the Spirit, the moment when frightened disciples become bold witnesses, and what many people consider the birthday of the Christian church.

Why This Matters
Pentecost reminds us that the Christian life isn't just about believing the right things – it's about being empowered by the Spirit to do the work of the Gospel. When AI helps you plan this service, you can create worship that recovers Pentecost's energy and commissions your congregation to live as Spirit-filled witnesses.

The Prompt
```
I'm planning a Pentecost worship service that
celebrates the Holy Spirit's outpouring and the
church's mission.

Create a worship service outline (60-75 minutes) that
includes:
- Opening that introduces Pentecost's significance
(many don't understand what we're celebrating)
- Scripture readings from the Pentecost narrative
(Acts 2:1-21, Joel 2:28-29, John 20:19-23, or Romans
8:22-27)
- 3-4 hymns focused on the Holy Spirit, wind, fire,
mission, and empowerment
- Sermon structure (15-20 minutes) on [theme: e.g.,
the Spirit empowers us for witness, Pentecost breaks
down barriers, we are God's sent people]
- A symbolic element (red paraments/stoles,
multilingual readings, wind/fire imagery,
commissioning ritual)
- Prayers for the Spirit's presence and boldness in
mission
```

```
- A benediction that commissions worshipers as
Spirit-filled witnesses

Worship style: [traditional/contemporary/blended]
Theological tradition: Wesleyan
Tone: Energetic, missional, globally minded
```

What to Expect
AI will generate a service outline with Pentecost-focused Scripture, hymn suggestions emphasizing the Spirit and mission, sermon structure, and ideas for symbolic elements. The service will help people understand what Pentecost means and what the Spirit empowers us to do.

Follow-Up Prompts
```
"Add a multilingual element where different people
read Acts 2 in different languages to embody
Pentecost's barrier-breaking power."
```
This makes the theological claim tangible – the Spirit enables communication and communion across human divisions.
```
"Suggest a commissioning ritual where worshipers
receive a specific mission challenge or are sent to
specific ministries."
```
Pentecost isn't just about celebration – it's about being sent. A ritual can make that sending concrete.
```
"Create a children's activity or object lesson using
wind or fire that teaches about the Holy Spirit's
power."
```
Kids can grasp Pentecost's energy through sensory, participatory experiences.

How to Use It
1. Use AI to create multilingual readings of Acts 2 for congregational participation
2. Have AI design a commissioning ritual sending people to specific ministries
3. Ask AI to create a children's activity using wind or fire to teach about the Spirit

Reflection Question
Where is the Holy Spirit calling your congregation to bold witness or risk-taking mission? What barriers – language, culture, fear, comfort – might the Spirit be asking you to cross?

CHAPTER 25
All Saints Day: Remembering the Faithful Departed

Our Goal
Create an All Saints Day service that honors our deceased loved ones, celebrates the communion of saints, and offers comfort to those grieving while affirming resurrection hope.

The Problem
All Saints Day walks the line between celebration and grief. You're honoring those who have died in the faith – both famous saints and beloved members of your congregation who've passed away. You want to create space for tears without making the service depressing, celebrating the saints' witness without ignoring the pain of loss.

Why This Matters
All Saints Day reminds us that we're part of a great cloud of witnesses stretching across time. The faithful who have gone before us still shape our faith. When AI helps you plan this service, you can create worship that holds grief and gratitude together, honoring the dead while affirming that death does not separate us from the love of God.

The Prompt
```
I'm planning an All Saints Day service that honors
our deceased loved ones, particularly church members
who have died in the past year.

Create a worship service outline (60-75 minutes) that
includes:
- Opening that explains the communion of saints and
why we remember those who have died
- Scripture readings on resurrection hope and the
cloud of witnesses (Revelation 7:9-17, Hebrews 11:1-
3, 12:1-2, John 11:32-44, or 1 John 3:1-3)
- 3-4 hymns about eternal life, the communion of
saints, and resurrection hope
- Sermon (12-15 minutes) on [theme: e.g., the
communion of saints, we are surrounded by witnesses,
death does not separate us from God's love]
- Remembrance ritual (reading names of the departed,
lighting candles, tolling a bell)
- Prayers for those grieving and thanksgiving for the
saints' witness
```

```
- Closing that affirms resurrection hope without
minimizing grief

Worship style: [traditional/contemporary/blended]
Theological tradition: Wesleyan
Tone: Reverent, comforting, hopeful but honest about
grief
```

What to Expect

AI will generate a complete service outline with Scripture focused on eternal life and the communion of saints, hymn suggestions, sermon structure, and detailed guidance for the remembrance ritual. The service will balance honoring the dead with comforting the living.

Follow-Up Prompts

```
"Write a script for the remembrance ritual where
names of the departed are read and candles are lit."
```
Clear liturgical structure helps the ritual feel reverent rather than awkward or rushed.
```
"Add a section that honors not just 'official' saints
but ordinary faithful people in our congregation."
```
All Saints Day isn't just about famous saints – it's about your church's faithful members who lived the Gospel quietly.
```
"Include a pastoral prayer that holds both grief and
resurrection hope without rushing past pain."
```
People need permission to grieve even as we affirm that death is not the end.

🫤 **Your Discernment Matters**: All Saints Day can be deeply emotional for those who've lost loved ones. Be pastorally present for grief. Don't force cheerfulness or rush to resurrection hope in ways that minimize pain. Create space for tears alongside affirmation of eternal life.

🔒 **Privacy Reminder**: If reading names of the departed, verify with families that they're comfortable with public acknowledgment. Some may prefer privacy in their grief.

How to Use It

1. Use AI to write a script for the remembrance ritual (reading names, lighting candles)
2. Have AI create a prayer that holds both grief *and* resurrection hope

3. Ask AI to adjust the service tone if your context calls for more celebration or more lament

Reflection Question

Who are the faithful witnesses in your life who have died? How does their example of faith continue to shape your discipleship?

CHAPTER 26
Christ the King Sunday: Reign of Christ

Our Goal
Create a Christ the King Sunday service that proclaims Jesus' reign over all creation while exploring what it means to live under His kingdom authority.

The Problem
Christ the King Sunday closes the liturgical year, but it's not widely recognized outside liturgical churches. The language of "king" and "kingdom" can feel archaic or troubling in our democracy-oriented culture. You need worship that proclaims God's sovereignty without sounding triumphalist, and that calls people to kingdom living without being heavy-handed.

Why This Matters
Christ the King Sunday declares that Jesus – not empire, nation, money, or power – has ultimate authority. This is a bold, countercultural claim. The service invites us to examine our true loyalties and reorient our lives around God's upside-down kingdom where the last are first and love triumphs over force.

The Prompt
```
I'm planning a Christ the King Sunday service that
proclaims Jesus' reign and explores what it means to
live under His kingdom authority.

Create a worship service outline (60-75 minutes) that
includes:
- Opening that introduces Christ the King Sunday
- Scripture readings on Christ's kingship (Jeremiah
23:1-6, Luke 23:33-43, Colossians 1:11-20, John
18:33-37, or Revelation 1:4b-8)
- 3-4 hymns focused on Christ's reign and kingdom
- Sermon structure (15-20 minutes) on [theme: e.g.,
what kind of king is Jesus, kingdom living, our
ultimate allegiance to Christ]
- Prayers acknowledging competing loyalties and
recommitting to Christ's authority
- A closing that transitions into Advent preparation

Worship style: [traditional/contemporary/blended]
Theological tradition: Wesleyan
```

```
Tone: Celebratory but not triumphalist, prophetic but
pastoral
```

What to Expect
AI will generate a service outline with Christ the King-focused Scripture, hymn suggestions emphasizing Jesus' lordship, sermon structure exploring kingdom themes, and prayers of commitment.

Follow-Up Prompts
```
"Add a section addressing why 'king' language might
be difficult and reframe it around Jesus' servant
leadership."
```
This acknowledges modern resistance to monarchical imagery while preserving the Biblical claim that Jesus has ultimate authority.
```
"Suggest contemporary worship songs that proclaim
Jesus' lordship without sounding triumphalist."
```
You can find music that celebrates Jesus' reign in ways that align with His upside-down kingdom values.
```
"Create a commissioning prayer that sends people to
live as kingdom citizens in their daily lives."
```
Christ the King Sunday should lead to concrete discipleship, not just theological affirmation.

🗿 **Your Discernment Matters**: "King" and "kingdom" language carries different weight in different cultural contexts. Be sensitive to those who've experienced oppressive political regimes. Ground Jesus' kingship in His servant leadership and self-giving love.

How to Use It
1. Adjust "king" language if your congregation needs gentler entry points to this imagery
2. Add specific examples of competing allegiances your congregation faces
3. Explicitly connect this service to Advent; this service ends the liturgical year and prepares us for Jesus' birth

Reflection Question
What competes most strongly for your ultimate allegiance besides Jesus, and what would it look like to truly let Jesus reign in that area of your life?

CHAPTER 27
Focusing on Social Justice: The UMC's Special Sundays

Our Goal
Create sermon content for the United Methodist Church's Special
Sundays: Human Relations Day, One Great Hour of Sharing, Native
American Ministries Sunday, Peace with Justice Sunday, World
Communion Sunday, and United Methodist Student Day.

The Problem
The UMC designates six Special Sundays throughout the year that focus
on justice, mission, and ministry. Each has its own emphasis and
offering. Creating unique content for six different Sundays feels
overwhelming when you're already managing weekly sermon prep and
pastoral care.

Why This Matters
These Special Sundays connect local congregations to global United
Methodist witness for justice and compassion. They fund crucial
ministries and educate congregations about issues many often ignore.
When AI helps you create sermon content for these days, you can preach
justice effectively without reinventing the wheel six times a year.

The Prompt
```
I'm a United Methodist pastor preparing a sermon for
[specific Special Sunday: Human Relations Day, One
Great Hour of Sharing, Native American Ministries
Sunday, Peace with Justice Sunday, World Communion
Sunday, or United Methodist Student Day].

Create sermon material with:
- Scripture passage connecting to this Sunday's
justice focus
- 3 main sermon points addressing Biblical
foundations
- Connection between Scripture and the UMC ministry
funded by this offering
- Application: here's what we can do
- Where the offering goes

Congregation: [size, demographics]
Theological tradition: Wesleyan social holiness
Tone: Prophetic but pastoral
Length: 300 words
```

What to Expect

AI will generate Scripture suggestions, sermon outlines, and connections between Biblical justice themes and the specific UMC ministry focus. Adaptable across all six Special Sundays by changing the ministry emphasis.

Follow-Up Prompts

"Add specific examples of how this UMC ministry has made a difference."
Concrete examples help people see real-world impact.
"Make the tone more hopeful and celebratory rather than guilt-focused."
Justice preaching should energize people for action.
"Create a brief prayer of commitment related to this Sunday's justice focus."
Worship elements beyond the sermon reinforce the emphasis.

Your Discernment Matters: Justice preaching can be politically divisive. Lead with Biblical foundations rather than partisan positions.

Scripture Check: Ensure Biblical texts actually support the justice claims you're making.

How to Use It

1. Customize your service and message with stories from actual UMC ministries funded by the offering
2. Add local connections showing how your congregation participates
3. Provide clear information about where contributions go

Reflection Question

Which of the six UMC Special Sundays most challenges your congregation's comfort zone and which most excites them?

SECTION 3

COMMUNICATION & OUTREACH

CHAPTER 28
Writing Church Newsletter Articles

Our Goal
Create engaging newsletter content that informs, inspires, and connects with your congregation without spending hours writing from scratch.

The Problem
Newsletters are essential for keeping your congregation informed and connected, but they're time-consuming to write. By the time you've finished your pastoral letter, event announcements, and ministry updates, you've burned through hours that could have been spent with people.

Why This Matters
Consistent, quality communication builds congregational connection and keeps people engaged in church life. When AI helps you draft newsletter content, you can maintain regular communication without sacrificing pastoral presence. Your congregation stays informed, and you stay sane.

The Prompt
```
I'm a [denomination] pastor writing our church
newsletter. Our congregation is [size, demographics,
location].

Create newsletter content for [month/season]
including:
1. A pastoral letter (150-200 words) on the theme of
[theme: e.g., gratitude, Advent preparation, summer
rest]
2. Three event announcements:
   - [Event 1: e.g., VBS, June 15-19]
   - [Event 2: e.g., Community Service Day, June 22]
   - [Event 3: e.g., Small Group Sign-Ups]
3. A brief spotlight (100 words) on [ministry or
volunteer]
4. A prayer or blessing related to [monthly holiday,
church event, or local event]

Tone: Warm, encouraging, conversational
Audience: Educated adults of all ages, primarily
[demographics]
```

What to Expect

AI will generate complete newsletter sections with appropriate length, tone, and structure. The pastoral letter will feel generic until you add personal details and stories from your ministry context.

Follow-Up Prompts

```
"Make the pastoral letter more concise – cut it to
100 words."
```

This teaches AI to adjust length while keeping your main message – useful anytime content feels too long.

```
"Rewrite the VBS announcement to appeal specifically
to parents of elementary-age children."
```

Once you have good content, you can ask AI to rewrite it for different audiences without starting over.

```
"Add a Scripture reference to the pastoral letter
that connects with the gratitude theme."
```

You can enrich any AI-generated content by asking it to add Biblical grounding.

😊 **Your Discernment Matters**: Generic AI-generated content can sound robotic. Always add personal stories, specific names, and local details that make the newsletter feel like it's from *your* church, not *any* church.

How to Use It

1. Use AI's draft as your starting framework and replace generic examples with specific church details
2. Ask AI to adjust tone or length for different newsletter sections as needed
3. Have AI generate multiple versions of announcements and pick the most engaging one
4. Ask AI to create a slide, min-poster, or flyer for one of your events based on the article you had it create.

Reflection Question

What recent church moment should be woven into this newsletter to make it feel authentically connected to your congregation's life together?

CHAPTER 29
Crafting Email Announcements for Events

Our Goal
Write clear, compelling email announcements that get people's attention, communicate essential information, and motivate action.

The Problem
Your inbox is full, and so is everyone else's. You need event announcements that cut through the noise – emails people actually open, read, and respond to. But crafting subject lines, writing engaging copy, and including all the details without overwhelming readers takes time you might not have.

Why This Matters
Email is still one of the most effective ways to reach your congregation, but only if people open it. A well-crafted announcement balances information with inspiration, logistics with invitation. When AI helps you draft these emails, you can communicate effectively without agonizing over every word.

The Prompt
```
I'm a [denomination] pastor writing an email
announcement for [event: e.g., VBS, mission trip,
capital campaign kickoff, new worship service].

Create an email announcement (200-250 words) that
includes:
- Compelling subject line with a relevant emoji to
catch people's attention
- Opening hook that grabs attention
- Clear event details (what, when, where, who, fees)
- Why this event matters
- Specific call to action (register, volunteer,
donate, attend)
- End with a warm closing urging readers to invite
their friends / spread the word

Event details: [provide specifics]
Audience: [congregation demographics]
Tone: [enthusiastic/reflective/urgent/invitational]
```

What to Expect
AI will generate a complete email with subject line, engaging opening, clear logistics, and call to action. The tone will match your request, though you'll need to personalize with local flavor.

Follow-Up Prompts
```
"Create 10 different subject lines with relevant
emojis so I pick which one I like the most."
```
You can pick what you think will resonate the most with your congregation.
```
"Make the call to action more urgent – we need
registrations by next Friday."
```
AI can adjust tone to create appropriate urgency without sounding manipulative.
```
"Rewrite this email for people who've never attended
this event before."
```
Different audiences need different framing – insiders vs. newcomers, participants vs. volunteers.

🙂 **Your Discernment Matters**: Generic AI announcements can miss the local flavor and inside references that make your church unique. Always personalize with specific details.

⚠ **Verify This**: Double-check all event details (dates, times, locations, registration links) before sending. AI-generated content can't verify factual accuracy.

How to Use It
1. Use AI's email as your foundation and add specific details AI couldn't know (deadline changes, personal endorsements)
2. Ask AI to convert someone's handwritten bullet points about an event into a compelling, informative message
3. Have AI create follow-up emails reminding people to register or volunteer

Reflection Question
What's the one action you most need people to take after reading this email, and is that action crystal clear?

CHAPTER 30
Creating Social Media Posts for Multiple Platforms

Our Goal
Generate engaging social media content for Facebook, Instagram, and other platforms that fits each platform's style while maintaining your church's voice.

The Problem
Each social media platform has its own culture, character limits, and best practices. What works on Facebook doesn't necessarily work on Instagram. You want to maintain an active social media presence, but creating platform-specific content for every post multiplies your workload.

Why This Matters
Social media is where many people – especially younger generations – discover churches and stay connected with their faith community. Consistent, engaging posts keep your church visible and accessible. When AI helps you create platform-specific content, you can maintain presence across multiple channels without spending hours on social media management.

The Prompt
```
I'm a [denomination] pastor creating social media
posts about [topic: e.g., upcoming sermon series,
community service event, inspirational message,
ministry highlight].

Create social media posts optimized for:
- Facebook (longer, conversational, link-friendly)
- Instagram (visual-focused, hashtags, emoji-
friendly)
- Twitter/X (concise, punchy, shareable)

Topic: [brief description]
Tone: [warm/inspirational/invitational/celebratory]
Include: Call to action where appropriate
Audience: [demographics and spiritual seekers]
```

What to Expect
AI will generate platform-specific versions of the same content, adjusted for each platform's style and character limits. You'll get different approaches rather than just truncated versions.

Follow-Up Prompts
"Add relevant hashtags to the Facebook post that connect with our local community."
Hashtags help new people discover your content, but they need to be relevant and authentic.
"Make the Facebook post more conversational – like I'm talking to a friend."
Tone adjustments help posts feel less corporate and more personal.
"Create an Instagram caption that pairs with the attached photo of our food pantry volunteers."
Visual content needs captions that enhance rather than just describe the image.

🌐 **Your Discernment Matters**: Social media posts represent your church to the public. Review AI-generated content for theological accuracy and appropriate tone before posting.

🔒 **Privacy Reminder**: Never share photos or information about individuals (especially children) on social media without explicit permission from adults or children's parents.

How to Use It
1. Use AI's platform-specific posts as starting points and add local details or timely references
2. Ask AI to adjust tone or length for different platforms based on what performs well with your audience
3. Have AI generate multiple post variations for A/B testing

Reflection Question
Which platform does your congregation engage with most, and how can you focus your energy there rather than trying to maintain equal presence everywhere?

CHAPTER 31
Developing a Social Media Content Calendar

Our Goal
Create a monthly social media content calendar that plans posts in advance, maintains consistent presence, and balances different types of content.

The Problem
Posting to social media reactively – whenever you remember or have a spare moment – leads to inconsistent presence and last-minute stress. You know you should plan ahead, but creating a month's worth of content ideas and scheduling them feels overwhelming.

Why This Matters
A content calendar transforms social media from a daily scramble into a strategic ministry tool. When you plan posts in advance, you ensure balanced content (not just event announcements), maintain consistent presence, and free yourself from daily content creation pressure. When AI helps you build the calendar, you can plan strategically without spending hours brainstorming post ideas.

The Prompt
```
I'm a [denomination] pastor creating a social media
content calendar for [month].

Generate a 30-day content calendar with post ideas
for [platforms: Facebook, Instagram, etc.] that
includes:
- Weekly themes (e.g., Scripture reflection Mondays,
ministry highlight Wednesdays, community engagement
Fridays)
- Mix of content types: inspirational quotes, event
announcements, behind-the-scenes, community stories,
Scripture reflections
- Specific post ideas for each day (brief
description, not full text)
- Balance between information and inspiration

Church context: [size, demographics, ministries]
Upcoming events: [list any major events to highlight]
Tone: [warm, invitational, authentic]
```

What to Expect
AI will generate a month-long calendar with daily post ideas organized by theme and content type. You'll get descriptions of posts, not full text – you can then use AI to write individual posts from the calendar.

Follow-Up Prompts
"Write the full post text for Monday Week 1's Scripture reflection idea."
Once you have the calendar framework, you can drill down and create actual posts for specific days.
"Adjust Week 3 to include more emphasis on our VBS registration deadline."
You can modify the calendar when priorities shift or deadlines approach.
"Add Instagram Story ideas for Sundays that encourage people to share their worship experience."
Different content formats can be layered into the calendar for varied engagement.

⚠ **Verify This**: AI can't know about late-breaking church events or last-minute schedule changes. Keep the posting calendar flexible and update it as circumstances change.

🧠 **Your Discernment Matters**: Balance promotional content with humor and spiritual insights or encouragement. A calendar that's all event announcements misses social media's community-building potential.

How to Use It
1. Use AI's calendar as your planning foundation and adjust post ideas based on what's actually happening in your church
2. Ask AI to write full posts for each calendar entry as you need them throughout the month
3. Have AI generate alternative post ideas if calendar suggestions don't fit your context

Reflection Question
What balance of inspirational content, event promotion, and community stories best serves your congregation's needs and reflects your church's personality?

CHAPTER 32
Writing Website Copy (About Us, Ministries, Visitor Info)

Our Goal
Create clear, welcoming website copy that tells your church's story, explains your ministries, and helps visitors know what to expect.

The Problem
Your church website is often the first impression people have of your congregation. Visitors are looking for basic information – who you are, what you believe, when you meet, what to expect – but they're also discerning whether this might be a spiritual home. Writing website copy that is both informative and invitational, clear and compelling, takes more time than you have.

Why This Matters
Website copy answers the questions visitors are too nervous to ask: "Will I fit here? What do they believe? Is this for people like me?" Well-written copy removes barriers and extends genuine welcome. When AI helps you draft website content, you can create pages that are both professional and pastoral.

The Prompt
```
I'm a [denomination] pastor writing website copy for
our [specific page: About Us, Ministries, What to
Expect, Beliefs, Visit Us].

Create website copy (200-300 words) that:
- Welcomes first-time visitors warmly
- Clearly communicates key information
- Reflects our church's personality and values
- Answers common visitor questions
- Includes a clear next step or invitation

Church context: [size, demographics, theological
tradition, location]
Target audience: [spiritual seekers, families, young
adults, etc.]
Tone: Warm, accessible, authentic (not churchy or
exclusive)
```

What to Expect
AI will generate website copy with welcoming tone, clear information, and appropriate structure. The content will be generic until you add specific details about your actual ministries, values, and community.

Follow-Up Prompts
```
"Make the 'What to Expect' page more specific about
logistics – parking, childcare, dress code, service
length."
```
Anxious visitors need concrete details, not just warm welcomes.
```
"Rewrite the 'About Us' page to emphasize our
commitment to justice and inclusion."
```
Your church's distinctive values and priorities should be clear and prominent.
```
"Create a brief FAQ section answering common first-
time visitor questions."
```
Anticipating and answering questions reduces anxiety and demonstrates hospitality.

👁 **Your Discernment Matters**: Website copy is often a visitor's first impression. Modify AI's language to capture your church's authentic personality and distinctive values.

✝ **Theological Caution**: Ensure your "Beliefs" or "What We Believe" page accurately reflects your denomination's theology and your congregation's actual convictions.

How to Use It
1. Use AI's copy as your foundation and infuse it with specific stories and examples from your actual church
2. Ask AI to adjust reading level or tone if the first draft doesn't match your congregation's voice
3. Have AI create multiple versions of key pages and test which language resonates most with visitors

Reflection Question
If a spiritual seeker visited your website right now, would they know they're genuinely welcome, or does your language assume insider knowledge?

CHAPTER 33
Drafting Visitor Welcome Emails

Our Goal
Create warm, personal welcome emails that make first-time visitors feel valued and encourage them to return without overwhelming them with information.

The Problem
First-time visitors fill out connection cards or sign guest books, and then... silence. You want to follow up quickly while they're still thinking about your church, but crafting individual welcome emails for every visitor takes time you don't have. Generic templates feel impersonal, but personalized emails for each guest are unsustainable.

Why This Matters
The follow-up email is your chance to say "we're genuinely glad you were here" and answer the questions visitors didn't ask on Sunday: What's next? How do I get connected? Will I fit here? A well-written welcome email can be the difference between someone returning and someone disappearing. When AI helps you draft these emails, you can send timely, warm follow-ups without sacrificing authenticity.

The Prompt
```
I'm a [denomination] pastor writing a welcome email
to a first-time visitor who attended worship on
[date].

Create a welcome email (150-200 words) that:
- Thanks them warmly for visiting
- Briefly mentions something specific about last
Sunday's service [theme, special music, sermon topic]
- Invites them to return and offers next steps
(coffee with pastor, small group, upcoming event)
- Provides key logistical info (service times,
childcare, website)
- Feels personal, not mass-produced

Church context: [size, demographics, worship style]
Tone: Warm, genuine, invitational (not pushy)
```

What to Expect
AI will generate a friendly welcome email with appropriate warmth and information. It will feel somewhat generic until you add the visitor's name and any specific details from your conversation with them.

Follow-Up Prompts
"Add a specific invitation to our upcoming new member orientation on [date]."
Concrete next steps help visitors move from "I visited once" to "I'm exploring membership."
"Make the tone more casual and conversational – less formal church language."
Your congregation's personality should come through in the welcome email's voice.
"Include a brief FAQ addressing common first-timer questions (parking, dress code, what to expect)."
Proactively answering anxieties removes barriers to returning.

🔒 **Privacy Reminder**: Only email people who gave you permission through connection cards or guest registration. Respect people's communication preferences and privacy.

👁 **Your Discernment Matters**: A welcome email that feels mass-produced or overly promotional can push visitors away. Personalize with genuine warmth.

How to Use It
1. Use AI's template and personalize with the visitor's name and any conversation details you remember
2. Ask AI to adjust the tone or emphasis based on what you know about the visitor (families, seekers, relocating members)
3. Have AI create different welcome email templates for different types of visitors – ex: first time, travelers, families, seniors, etc.

Reflection Question
What one thing could you include in this email that would make the visitor feel genuinely seen and valued, not just processed?

CHAPTER 34
Creating Multi-Week Follow-Up Sequences
to Welcome New Members

Our Goal
Design a multi-week email sequence that nurtures new members through their first months, helping them get connected and feel at home.

The Problem
New members join your church excited and hopeful, and then... they drift often away because they don't know how to get involved, meet people, or find their place. You want to walk with new members through their first 90 days, but creating a consistent follow-up system feels overwhelming.

Why This Matters
The first few months determine whether new members stick or disappear. A planned follow-up sequence ensures no one falls through the cracks. Each email can introduce a ministry, invite connection, or answer common questions. When AI helps you create this sequence, you can provide consistent pastoral care to new members without starting from scratch each time.

The Prompt
```
I'm a [denomination] pastor creating a 6-week email
follow-up sequence for new members.

Create a sequence with one email per week that:
Week 1: Welcome and affirm their membership decision
Week 2: Introduce small groups and connection
opportunities
Week 3: Explain worship and spiritual formation
practices
Week 4: Highlight service and mission opportunities
Week 5: Introduce church governance and how to get
involved in leadership
Week 6: Invite ongoing discipleship and next steps

Each email: 150-200 words
Tone: Warm, encouraging, informational
Church context: [size, demographics, ministries]
```

What to Expect
AI will generate six complete emails, one for each week, with appropriate themes and progression. The sequence will feel structured but may need personalization for your specific ministries and culture.

Follow-Up Prompts
"Adjust Week 2 to emphasize our [specific small group format or discipleship pathway]."
Generic references to "small groups" should become specific invitations to actual groups.
"Add a personal story or testimony in Week 4 about someone who found their place through mission work."
Real stories make abstract invitations concrete and compelling.
"Create a shorter 4-week sequence instead of 6 weeks."
Some congregations need faster onboarding; AI can condense the sequence.

⚙ **Your Discernment Matters**: Not all new members need the same onboarding journey. Be prepared to adjust the sequence based on individual needs and maturity levels.

⚠ **Verify This**: Update ministry contact information and event details regularly. An email link pointing to outdated information frustrates rather than helps.

How to Use It
1. Use AI's sequence as your framework and customize each email with actual ministry names and contact people
2. Ask AI to adjust timing or themes based on your church's specific onboarding process
3. Have AI create branch sequences for different new member groups (families, young adults, retirees)

Reflection Question
What's the most common reason new members drift away from your church, and does this email sequence address that specific barrier?

CHAPTER 35
Writing Thank-You Notes to Volunteers and Donors

Our Goal
Create heartfelt, specific thank-you notes that genuinely honor volunteers and donors without sounding generic or obligatory.

The Problem
Your volunteers and donors deserve sincere gratitude, but writing dozens of personalized thank-you notes takes hours you might not have. Generic "thank you for your service" notes feel hollow. You want to honor people's generosity and hard work with words that actually mean something.

Why This Matters
Gratitude builds culture. When people feel genuinely seen and appreciated, they're more likely to continue serving and giving. A well-written thank-you note validates someone's contribution and reinforces that their effort matters. When AI helps you draft these notes, you can express authentic gratitude efficiently.

The Prompt
```
I'm a [denomination] pastor writing a thank-you note
to [volunteer/donor] who [specific contribution:
e.g., led VBS, served in food pantry, gave generously
to building fund].

Create a thank-you note (100-150 words) that:
- Expresses sincere gratitude
- Names the specific contribution or gift
- Describes the impact of their service/generosity
- Feels personal, not mass-produced
- Closes warmly

Tone: Warm, genuine, honoring
Relationship: [long-time member, new volunteer,
anonymous donor, etc.]
```

What to Expect
AI will generate a complete thank-you note with appropriate warmth and specificity. It will feel generic until you add personal details about the individual and their unique contribution.

Follow-Up Prompts

```
"Make this more specific about how their food pantry
work helped the Martinez family get through a hard
month."
```

Concrete impact stories make gratitude feel real rather than abstract.

```
"Adjust the tone to be more casual and friendly –
this volunteer is a close friend."
```

Different relationships warrant different levels of formality.

```
"Add a sentence inviting them to serve again at our
next event."
```

Gratitude can naturally lead to future involvement without feeling manipulative.

😊 **Your Discernment Matters**: Generic gratitude feels hollow. AI can't know the specific sacrifice or faithfulness behind someone's service – you must add that.

🔒 **Privacy Reminder**: When thanking donors, respect their privacy. Don't publicly share giving amounts or financial details without explicit permission.

How to Use It

1. Use AI's note as your foundation and add specific details about the person and their contribution
2. Ask AI to adjust tone based on your relationship with the recipient
3. Have AI create thank-you note templates for recurring volunteer roles (ex: food pantry) or donation types (ex: new roof fund)

Reflection Question

What specific detail about this person's service or gift could you mention that would let them know you genuinely noticed, not just sent an automatic thank-you?

CHAPTER 36
Crafting Press Releases for Community Events

Our Goal
Write professional press releases that get local media attention for your church's community events, outreach initiatives, or newsworthy activities.

The Problem
You're hosting a community event that could genuinely serve your neighborhood, but local media won't cover it unless you send a proper press release. Writing press releases requires a specific format and journalistic style you may not know. Poor press releases get ignored; good ones get coverage.

Why This Matters
Local media coverage extends your reach beyond your congregation into the broader community. A well-written press release can bring new people to your events, raise awareness of your ministries, and position your church as a community partner. When AI helps you write press releases, you can communicate professionally with media outlets without learning journalistic conventions from scratch.

The Prompt
```
I'm a [denomination] pastor writing a press release
for [event: e.g., free community meal, back-to-school
supply drive, affordable housing partnership,
interfaith service].

Create a press release (250-300 words) that follows
journalism standards:
- Compelling headline
- Strong opening paragraph with who, what, when,
where, why
- Quote from church leader (me) explaining
significance
- Event details including cost(s) [provide any fees],
community impact, and how people can sign up or get
involved
- Contact information for media inquiries
- Closing boilerplate about our church

Event details: [provide specifics]
```

```
Community benefit: [who this serves and why it
matters]
Tone: Professional, newsworthy, community-focused
```

What to Expect
AI will generate a properly formatted press release with headline,
inverted pyramid structure, quote, and boilerplate. The content will be
professional and suitable for media submission.

Follow-Up Prompts
```
"Make the headline more attention-grabbing –
emphasize the free aspect and community need."
```
Headlines determine whether journalists keep reading, so they need to be
compelling.
```
"Add statistics about food insecurity in our county
to strengthen the 'why this matters' section."
```
Data makes the community need concrete and newsworthy.
```
"Adjust the quote to emphasize our church's
partnership with other community organizations."
```
Collaborative efforts are often more newsworthy than solo church events.

⚠ **Verify This**: Make sure all facts, statistics, dates, and contact
information in press releases are accurate. Journalists won't fact-check
for you.

🌐 **Your Discernment Matters**: What's newsworthy to you may not be
newsworthy to media outlets. Focus on genuine community impact, not
just church promotion.

How to Use It
1. Use AI's press release and add specific local data or partnership
   details
2. Ask AI to adjust the angle or emphasis based on what aspects are
   most newsworthy
3. Have AI create different versions targeting different media
   outlets (newspaper, radio, community blog)

Reflection Question
What makes this event genuinely newsworthy to people outside your
church, and is that angle clear in the first paragraph?

CHAPTER 37
Developing Sermon Recap Emails for Monday

Our Goal
Create Monday emails for your congregation that recap Sunday's sermon, offer reflection questions, and extend pastoral care into the week.

The Problem
Sunday's sermon ends when people leave the sanctuary, but you want the message to continue shaping their week. A Monday recap email keeps the sermon alive, but writing these emails every week adds another task to your already busy Monday workload.

Why This Matters
Monday morning is when people return to work, family demands, and the ordinary rhythms of life. A sermon recap email meets them there, reminding them that Sunday's message applies to Monday's challenges. When AI helps you create these recaps, you can extend your pastoral presence into the week.

The Prompt
```
I preached a sermon on [Scripture passage] titled
"[sermon title]" with these main points:
1. [Point 1]
2. [Point 2]
3. [Point 3]

[If possible, attach a PDF of your sermon script.]

Create a Monday morning email (200-250 words) that:
- Briefly recaps the sermon's main message
- Offers 2-3 reflection questions for the week ahead
- Includes a short prayer or blessing related to the
theme
- Encourages application to daily life
- Warm, pastoral tone (not preachy recap)

Congregation: [demographics]
Tone: Encouraging, accessible, Monday-morning
appropriate
```

What to Expect

AI will generate a complete recap email with summary, questions, and prayer. The content will feel somewhat generic until you personalize it with specific examples or stories from Sunday's actual service.

Follow-Up Prompts

"Make the reflection questions more specific to workplace challenges – many in our congregation are navigating difficult work environments."

Generic questions become useful when they address your congregation's actual lived experience.

"Add a brief story or illustration that reinforces the sermon's main point."

Sometimes Monday needs a fresh angle on Sunday's message rather than just repetition.

"Create a short summary of the message (100 words) for people who prefer quick reads."

Busy people appreciate concise emails that respect their time.

Your Discernment Matters: AI can't capture the Spirit-led moments in your actual Sunday service or spontaneous conversations from Coffee Hour afterwards. Add those authentic details yourself.

Verify This: Make sure reflection questions and applications don't contradict what you actually said in the sermon. AI might interpret your points differently.

How to Use It

1. Use AI's recap as your foundation and add a personal greeting or detail from Sunday's service
2. Ask AI to adjust the reflection questions based on feedback from parishioners after the service
3. Have AI create different recap formats (devotional style, action-focused, contemplative) and test what resonates

Reflection Question

What one question could help your congregation carry Sunday's message into Monday's real-world challenges?

CHAPTER 38
Repurposing Sermons into Blog Posts

Our Goal
Transform your Sunday sermon into a blog post that reaches people beyond your sanctuary and extends your teaching ministry online.

The Problem
You've invested hours preparing your sermon, but it reaches only the people who attended Sunday worship. A blog post could extend that message to visitors checking out your website, members who missed church, and people searching online for spiritual content. But rewriting sermons as blog posts feels like creating entirely new content.

Why This Matters
Blog posts give your sermons a second life. They're searchable, shareable, and accessible to people who will never walk through your church doors. A well-written blog post can introduce spiritual seekers to your teaching, reinforce Sunday's message for members, and establish your church's online presence. When AI helps you repurpose sermons, you multiply your ministry impact without multiplying your workload.

The Prompt
```
I preached a sermon on [Scripture passage] titled
"[sermon title]" with these main points:
1. [Point 1]
2. [Point 2]
3. [Point 3]

[If possible, attach a PDF of your sermon script.]

Repurpose this sermon into a blog post (400-500
words) that:
- Opens with a compelling hook or question
- Summarizes the main Biblical message
- Offers practical application
- Ends with invitation to reflection or action
- Reads naturally as written content, not a
transcript

Audience: Website visitors and church members
Tone: Conversational, accessible, pastoral
```

What to Expect

AI will generate a complete blog post that captures your sermon's essence in written form. It will feel more polished and readable than a sermon transcript but may need your personal voice and specific stories.

Follow-Up Prompts

```
"Add a personal story or illustration from my actual
sermon that makes the point more concrete."
```
Your real sermon illustrations make the blog post feel authentic rather than generic.
```
"Make the opening more attention-grabbing - start
with a question or surprising statement."
```
Blog readers skim, so strong openings are crucial for keeping their attention.
```
"Adjust this for spiritual seekers who may not be
familiar with church language."
```
Blog posts reach beyond your congregation, so accessibility matters.

📖 **Scripture Check**: When AI converts sermon content to blog format, verify that Biblical references and theological claims remain accurate and in context.

🧠 **Your Discernment Matters**: Blog posts reach people outside your congregation who don't know you. Ensure the tone and content work for spiritual seekers, not just church insiders.

How to Use It

1. Use AI's blog post as your foundation and infuse it with your actual sermon stories and examples
2. Ask AI to adjust length or reading level based on your website audience
3. Have AI create SEO-friendly titles, hashtags, and meta descriptions that help people find the post

Reflection Question

What's one insight from your sermon that could genuinely help someone who's never set foot in your church?

CHAPTER 39
Creating Memes from Sermon Content

Our Goal
Design simple, shareable memes that capture key sermon insights in visual, social-media-friendly format.

The Problem
Memes are how ideas spread online, especially among younger generations. You want to share sermon insights on social media in ways that actually get attention and shares, but you're not a graphic designer who can distill complex theology into meme format.

Why This Matters
A well-crafted meme can take your sermon's key insight and put it in front of hundreds or thousands of people who would never read a blog post or watch a sermon video. Memes are shareable, memorable, and accessible. When AI helps you create meme concepts, you can meet people where they already are – scrolling social media.

The Prompt
```
I preached a sermon on [Scripture passage] with the
key message: [one-sentence summary of main point].

[If possible, attach a PDF of your sermon script.]

Create 3-5 meme concepts that:
- Capture the sermon's main insight in 10-20 words
- Use accessible, non-churchy language
- Are relatable to everyday life
- Work well with simple visual imagery
- Are shareable and memorable

Format: Provide the text for each meme and suggest a
simple image concept (photo, graphic, emoji)
Tone: Conversational, relatable, warm
Audience: Social media users, including spiritual
seekers
```

What to Expect
AI will generate multiple meme concepts with text and image suggestions. You'll still need to create the actual graphics using Canva, Meme Generator, or similar tools, but AI provides the content strategy.

Follow-Up Prompts
"Make meme #2 more humorous – it's too serious for meme format."
Memes thrive on relatability and often gentle humor, not heavy theology.
"Suggest specific free stock photo sites where I could find an image matching meme concept #3."
AI can point you toward image resources that match the meme concept.
"Create a version of meme #1 that references a current cultural moment or trend."
Timely memes get more engagement than generic ones.

🫥 **Your Discernment Matters**: Memes trivialize complex theology if not handled carefully. Use memes for accessible entry points to deeper truth, not as replacements for substantive teaching.

✝️ **Theological Caution**: Ensure meme content doesn't oversimplify or misrepresent Biblical truth for the sake of being shareable or humorous.

How to Use It
1. Use AI's meme concepts and create actual graphics using free tools like Canva
2. Be an Art Director: Ask AI to generate multiple versions based on different themes and test which ones get the most engagement
3. Have AI adapt meme concepts for different platforms (Instagram, Facebook, Twitter)

Reflection Question
What's the one insight from your sermon that would actually make someone stop scrolling and think?

CHAPTER 40
Writing Video Scripts for Ministry Highlights

Our Goal
Create short video scripts that showcase your church's ministries, tell impact stories, and invite people into your community.

The Problem
Video is the most engaging content format online, but writing scripts for ministry highlight videos takes time and skill you may not have. You want to capture your church's personality and mission in 60-90 seconds, but translating ministry impact into compelling video narrative is harder than it looks.

Why This Matters
Video tells stories in ways text and photos cannot. A well-scripted ministry highlight video can help visitors understand who you are, inspire members to get involved, and share your church's mission with the broader community. When AI helps you write video scripts, you can create engaging content without becoming a professional videographer.

The Prompt
```
I'm creating a video highlighting [ministry: e.g.,
food pantry, youth program, mission trip, worship
experience].

Write a video script (60-90 seconds when spoken)
that:
- Opens with a compelling hook or question
- Introduces the ministry and its purpose
- Shares a brief impact story or testimony
- Explains how people can get involved or support
- Closes with invitation or call to action

Include: Suggested visuals or B-roll descriptions
Tone: Warm, authentic, inspiring (not overly
promotional)
Audience: Website visitors, social media viewers,
potential volunteers
```

What to Expect
AI will generate a complete video script with dialogue, visual suggestions, and timing. The script will provide structure, though you'll

need to personalize with actual stories and testimonies from your ministry.

Follow-Up Prompts
```
"Add this specific testimony from Maria who
volunteers in our food pantry every Tuesday. [Insert
the quote.]"
```
Real names and stories make video scripts authentic rather than promotional.
```
"Make the opening more attention-grabbing – we need
to hook viewers in the first 5 seconds."
```
Online video viewers decide whether to keep watching almost immediately.
```
"Create a shorter 30-second version for Instagram
Reels or TikTok."
```
Different platforms require different video lengths and pacing.

🔒 **Privacy Reminder**: Get written permission before featuring anyone in ministry videos, especially children. Not everyone wants to be on camera representing your church.

😌 **Your Discernment Matters**: Video scripts should tell authentic stories, not create promotional marketing that feels manipulative or exaggerated.

How to Use It
1. Use AI's script as your foundation and insert actual testimonies and ministry details
2. Ask AI to adjust pacing or emphasis based on your video platform and audience
3. Have AI create multiple script versions for different ministries or campaigns

Reflection Question
What's the one story from your church's ministries that would help someone understand why that ministry matters and want to be part of it?

CHAPTER 41
Drafting Photo Captions for Social Media

Our Goal
Write engaging photo captions that provide context, tell stories, and encourage interaction on social media.

The Problem
You have great photos from Sunday worship, mission trips, or church events, but posting them without compelling captions wastes their potential. Good captions turn photos from "here's what happened" into stories that connect, inspire, and invite engagement. But writing fresh, engaging captions for every photo takes creative energy you're short on.

Why This Matters
Photos get attention, but captions create connection. A well-written caption gives context, invites response, and turns a static image into a conversation starter. When AI helps you write captions, you can post consistently engaging content without caption-writing burnout.

The Prompt
```
I'm posting a photo or illustration on [platform:
Facebook, Instagram] showing [describe the image:
e.g., volunteers serving at food pantry, children
during VBS, congregation worshiping, mission trip
team].

[If possible, attach the photo or illustration.]

Write a caption (50-150 words depending on platform)
that:
- Provides context for the photo
- Tells a brief story or highlights impact
- Includes a question or invitation to engage
- Uses appropriate hashtags (for Instagram)
- Matches our church's voice and values

Tone: [warm/celebratory/reflective/inspirational]
Audience: Church members and community followers
```

What to Expect
AI will generate platform-appropriate captions with context, storytelling,
and engagement prompts. Instagram captions will include hashtag
suggestions; Facebook captions will be more conversational.

Follow-Up Prompts
```
"Make this caption more personal - include the name
of the volunteer pictured and what motivates her
service."
```
Naming real people and their stories makes captions feel authentic.
```
"Add a call to action inviting people to volunteer at
our next food pantry day."
```
Captions can inspire action, not just likes.
```
"Create an Instagram Story caption version - shorter
and more casual."
```
Stories require different caption style than feed posts.

🔒 **Privacy Reminder**: Never post photos of minors without explicit
parental permission. Some families don't want their children's images
online. Privacy is vital for people who are being harassed.

🫣 **Your Discernment Matters**: Captions should honor the people in the
photo. Avoid making jokes or comments that could embarrass or
misrepresent them.

How to Use It
1. Use AI's caption as your starting point and personalize with
   names and specific details
2. Ask AI to adjust tone or length based on the platform and photo
   content
3. Have AI generate multiple caption options and pick the most
   engaging one

Reflection Question
What's happening in this photo that tells a bigger story about who your
church is and what you value?

CHAPTER 42
Creating Email Campaigns for Stewardship or Capital Campaigns

Our Goal
Design a multi-email campaign that educates, inspires, and invites generosity during stewardship seasons or capital campaigns.

The Problem
Asking for money is uncomfortable, but funding ministry is essential. You need a stewardship campaign that honors people's generosity, casts vision for ministry impact, and invites financial commitment without manipulation or guilt. Writing multiple emails that build momentum and tell a compelling story takes significant time and communication skill.

Why This Matters
Stewardship campaigns shape your congregation's understanding of generosity and fund the ministries that change lives. A well-designed email campaign educates givers about ministry impact, invites them into partnership, and normalizes faithful financial stewardship. When AI helps you create the campaign, you can communicate effectively without spending weeks on email drafts.

The Prompt
```
I'm creating a [4-week/6-week] email campaign for
[annual stewardship/building campaign/mission
fundraiser].

Create a campaign with one email per week that:
Week 1: Cast vision - why this campaign matters
Week 2: Share impact stories - how previous giving
made a difference
Week 3: Explain the need - specific financial goals
and ministry plans
Week 4: Invite commitment - clear ask and how to give

Each email: 200-250 words
Include: Subject lines, specific calls to action
Tone: Inspiring, grateful, invitational (not guilt-
driven)
Campaign goal: [dollar amount and purpose]
Church context: [size, giving culture, ministries
funded]
```

What to Expect
AI will generate a complete multi-week campaign with emails, subject lines, and calls to action. The progression will build logically from vision to invitation, though you'll need to personalize with actual ministry stories and financial details.

Follow-Up Prompts
"Add a testimony in Week 2 from someone whose life was changed by our food pantry ministry. [Insert their story and quotes.]"
Real stories make the case for giving more compelling than abstract appeals.
"Adjust Week 3 to include specific budget details showing where contributions actually go."
Transparency builds trust and helps people understand their giving's impact.
"Create a final reminder email for people who haven't yet pledged."
Campaigns often need a gentle follow-up for those who intended to respond but haven't yet.

🫥 **Your Discernment Matters**: Fundraising appeals can easily slip into manipulation or guilt tactics. Ensure your campaign invites generous partnership, not coerced giving.

⚠ **Verify This**: All financial goals, budget details, and ministry impact claims must be accurate and transparent. Trust is built on honesty.

How to Use It
1. Use AI's campaign structure and insert actual ministry impact stories and financial details
2. Ask AI to adjust timing or emphasis based on your congregation's giving patterns
3. Have AI create variations for different donor segments (first-time givers, major donors, sustaining members)

Reflection Question
What ministry impact story would most help your congregation see their giving as partnership in God's work rather than institutional obligation?

CHAPTER 43
Writing Fundraising Appeal Letters

Our Goal
Create compelling fundraising appeal letters that tell stories, cast vision, and invite generous giving for specific ministry needs or special projects.

The Problem
Fundraising letters walk a fine line between inspiring generosity and feeling manipulative. You need to communicate urgent ministry needs, celebrate past impact, and invite financial partnership – all while honoring donors and avoiding guilt tactics. Writing letters that accomplish all of this takes skill and time you're short on.

Why This Matters
Well-written fundraising appeals fund ministries that change lives. They educate donors about impact, invite them into partnership with God's work, and normalize generous giving. When AI helps you draft these letters, you can communicate effectively about financial needs without spending days on a single appeal.

The Prompt
```
I'm writing a fundraising appeal letter for [specific
need: mission trip, building repair, scholarship
fund, ministry expansion].

Create a letter (300-400 words) that:
- Opens with a compelling story or need
- Explains the ministry's impact and importance
- States the specific financial goal clearly
- Shows how donations will be used
- Includes a clear call to action and giving
instructions
- Closes with gratitude and vision

Audience: Current donors and church members
Tone: Grateful, inspiring, urgent but not
manipulative
Ministry context: [describe the specific need and
impact]
```

What to Expect
AI will generate a complete fundraising letter with story, vision, financial ask, and call to action. The letter will be professionally structured but generic until you personalize with actual ministry stories and donor relationships.

Follow-Up Prompts
"Add a specific story about how last year's mission trip changed Maria's life."
Real stories make abstract ministry needs concrete and compelling.
"Make the financial ask more specific – we need $15,000 by June 30th for the roof repair."
Specific goals and deadlines create urgency and clarity.
"Adjust the tone to be more personal and less formal – this is going to longtime supporters who know us well."
Different donor relationships warrant different levels of formality.

💭 **Your Discernment Matters**: Fundraising stories should be true and honor the dignity of those receiving ministry. Never exploit people's struggles for donation appeals.

⚠ **Verify This**: Ensure all claims about how donations will be used are accurate. Donors deserve transparency about where their money goes.

🔒 **Privacy Reminder**: Get permission before sharing anyone's personal story in a fundraising appeal, even if the story is inspiring.

How to Use It
1. Use AI's letter structure and insert actual ministry impact stories and donor testimonies
2. Ask AI to adjust the urgency level based on your timeline and fundraising need
3. Have AI create different letter versions for different donor segments (major donors, first-time givers, monthly supporters)

Reflection Question
What ministry impact would most help donors see their giving as investment in God's kingdom rather than just meeting a budget need?

CHAPTER 44
Developing Church Branding and Taglines

Our Goal
Create clear, memorable church branding language and taglines that communicate your identity, values, and mission in just a few words.

The Problem
Your church has a mission statement buried somewhere in official documents, but you need something shorter and more memorable for your website, social media, and outreach materials. Distilling your church's identity into a compelling tagline that's both accurate and engaging is harder than it looks.

Why This Matters
Good branding helps people quickly understand who you are and whether they might belong. A clear tagline answers the unspoken question every visitor has: "What's this church about?" When AI helps you develop branding language, you can articulate your identity clearly without hiring expensive consultants.

The Prompt
```
I'm developing branding language for our church that
captures our identity and mission.

Create branding options including:
- 3-5 tagline options (5-10 words each)
- A brief mission statement (one sentence)
- Core values statements (3-5 values with brief
explanations)

Our church is characterized by: [theological
tradition, ministry focus, community context,
distinctive values]
We want people to know we're: [welcoming, justice-
focused, traditional, contemporary, family-oriented,
etc.]
Tone: Clear, memorable, authentic (not trendy or
corporate)
```

What to Expect

AI will generate multiple tagline options, mission statement drafts, and core values language. The options will give you starting points to refine, combine, or adapt to your actual church culture.

Follow-Up Prompts

```
"Make tagline option #2 more specific about our
commitment to racial justice and inclusion."
```
Generic values need specificity to be meaningful and memorable.
```
"Create a version that emphasizes our
intergenerational community rather than just 'all are
welcome.'"
```
What makes your church distinctive should come through in your branding.
```
"Combine elements from tagline #1 and #3 into a
single stronger option."
```
Sometimes the best tagline emerges from blending multiple AI-generated options.

🧠 **Your Discernment Matters**: Your church's identity should be defined by your actual community and values, not by trendy branding language AI generates.

✝ **Theological Caution**: Ensure branding language accurately reflects your theological tradition and denominational commitments, not just marketable buzzwords.

How to Use It

1. Use AI's tagline options as starting points and test them with church leadership and members
2. Ask AI to refine promising options based on feedback about what resonates
3. Have AI develop supporting branding language once you've settled on a tagline

Reflection Question

If someone asked a member of your congregation "What's your church about?" what would you want them to be able to say in one sentence?

CHAPTER 45
Creating Podcast Episode Outlines from Sermon Series

Our Goal
Develop podcast episode outlines that extend your sermon series into ongoing discipleship content, reaching people beyond Sunday morning.

The Problem
Podcasting could expand your teaching ministry beyond your sanctuary, but creating original podcast content on top of sermon prep feels impossible. You want to leverage your sermon series without just uploading sermon audio and calling it a podcast.

Why This Matters
Podcasts meet people where they are – commuting, exercising, doing chores. A well-designed podcast based on your sermon series extends your teaching ministry, provides midweek discipleship, and reaches people who might never attend Sunday worship. When AI helps you create podcast outlines, you can launch a podcast without doubling your content creation workload.

The Prompt
```
I'm creating a podcast based on my sermon series on
[topic/Scripture book]. The series has [number]
sermons with these themes:
1. [Theme 1]
2. [Theme 2]
3. [Theme 3]

[If possible, attach a PDF of your sermon script.]

Create podcast episode outlines (one per sermon
theme) that:
- Include episode title and description
- Suggest discussion format (solo teaching,
interview, Q&A, conversational)
- Outline 3-5 main segments or talking points
- Include questions for listener engagement or
reflection
- Suggest episode length (20-30 minutes)

Podcast audience: Church members and spiritual
seekers
```

```
Tone: Conversational, accessible, engaging (less
formal than sermons)
```

What to Expect
AI will generate episode outlines with titles, format suggestions, segment breakdowns, and discussion prompts. The outlines provide structure for recording but leave room for your personality and spontaneity.

Follow-Up Prompts
```
"Add interview question ideas for Episode 2 - I want
to bring in a guest who works in social justice."
```
Guest interviews add variety and expertise to podcast content.
```
"Create listener Q&A prompts I could use to
crowdsource questions for Episode 4."
```
Audience engagement makes podcasts feel interactive rather than one-directional.
```
"Suggest show notes and resource links I should
include for Episode 1."
```
Good show notes add value and help listeners go deeper into topics.

📖 **Scripture Check**: Podcast formats allow longer discussion, but verify that extended theological reflection remains Biblically grounded and contextually accurate.

🧠 **Your Discernment Matters**: Podcasts reach beyond your congregation. Consider how your teaching will be received by listeners who don't know your church or context. Avoid or explain churchy language and abbreviations like NT, UMC, or PC-USA.

How to Use It
1. Use AI's episode outlines as your recording framework and add personal stories and current examples
2. Ask AI to suggest different episode formats (solo, interview, panel) to explore options and create variety
3. Have AI create promotional social media posts for each episode

Reflection Question
What aspect of your sermon series would benefit from the longer, more conversational format podcasting allows? Where would doing a deep dive help listeners better understand the Scripture?

SECTION 4

PASTORAL CARE & COUNSELING

CHAPTER 46
Preparing for Hospital Visits

Our Goal
Research medical conditions and prepare for pastoral hospital visits so you can offer informed, compassionate care without saying something inappropriate or hurtful.

The Problem
When a church member is hospitalized, you want to visit promptly and offer meaningful pastoral care. But you may not understand their medical condition, prognosis, or what they're experiencing physically and emotionally. Walking into a hospital room unprepared can lead to awkward silences, insensitive comments, or missed opportunities for genuine comfort.

Why This Matters
Hospital visits are sacred pastoral moments. People are vulnerable, scared, and often isolated. A well-prepared visit demonstrates care and competence. When AI helps you research medical conditions and plan visits, you can arrive ready to offer the presence and prayer people need.

The Prompt
```
I'm visiting a church member hospitalized for
[medical condition: e.g., heart surgery, cancer
treatment, stroke recovery, chronic illness flare-
up].

Help me prepare by providing:
- Basic explanation of the condition in non-medical
language
- What the patient might be experiencing physically
and emotionally
- What to expect during recovery or treatment
- Appropriate things to say or ask
- Pastoral care considerations (when to pray, how
long to stay, sensitivity to family dynamics)

Context: [patient's age, family situation, church
involvement]
Length: 200-300 words
Tone: Informative, pastorally sensitive
```

What to Expect
AI will provide clear medical information, emotional context, and pastoral guidance. The information helps you understand what the patient is facing without needing medical school.

Follow-Up Prompts
"What specific prayers or Scripture passages would be most comforting for someone facing this diagnosis?"
Matching your pastoral care to the patient's actual experience makes visits more meaningful.
"What questions should I avoid asking that might seem insensitive or intrusive?"
Knowing what NOT to say is as important as knowing what to say.
"How can I best support the family members who are also struggling with this diagnosis?"
Hospital visits often involve caring for caregivers, not just patients.

🔒 **Privacy Reminder**: Never input a patient's actual name or identifying details into AI when researching their condition. Use general descriptions only.

⚠ **Verify This**: AI provides general medical information, not professional medical advice. **Do not share medical opinions or treatment suggestions** with patients or families.

🕊 **Your Discernment Matters**: Every patient and situation is unique. Use AI's guidance as background preparation, but let the actual visit be guided by the Spirit and the patient's immediate needs.

How to Use It
1. Use AI to research the medical condition before the visit so you understand what the patient is experiencing
2. Ask AI for pastoral care suggestions but adapt them based on your actual relationship with the patient
3. Have AI prepare appropriate follow-up care plans (prayer lists, meal trains, ongoing support) after the visit

Reflection Question
How can you balance being informed about their medical situation with giving them space to share what they want you to know?

CHAPTER 47
Writing Condolence Letters and Sympathy Cards

Our Goal
Craft heartfelt, pastoral condolence letters that offer comfort and acknowledge grief without sounding generic or impersonal.

The Problem
When someone in your congregation experiences loss, you want to reach out quickly with words of comfort. But grief is sacred and personal – you can't afford to sound robotic or use clichés. Finding the right words when your own heart is heavy (and your schedule is full) is one of pastoral ministry's most emotionally draining tasks.

Why This Matters
A well-written condolence letter is a ministry of presence when you can't physically be there. It assures grieving families that they're not alone, that their loved one mattered, and that their church family surrounds them in prayer. Getting the tone right matters deeply.

The Prompt
```
I'm a [denomination] pastor writing a condolence
letter to [name/family name] who recently lost
[relationship: e.g., their mother, their spouse,
their child].

Write a pastoral condolence letter (150-200 words)
that:
- Acknowledges the loss with appropriate gravity
- Offers comfort without clichés or easy answers
- Includes a brief, appropriate Scripture reference
or spiritual truth
- Assures them of prayer and church support
- Closes with warmth and availability

Tone: Compassionate, genuine, pastoral
Avoid: Phrases like "they're in a better place," "God
needed another angel," or "everything happens for a
reason"
Theological perspective: [Your tradition's approach
to grief and hope]
```

What to Expect
AI will generate a structurally sound letter with compassionate language.
It will feel somewhat generic until you add personal details about the
deceased and your relationship with the family.

Follow-Up Prompts
```
"Make this letter more personal by referencing a
specific memory or quality of the deceased."
```
Adding specific details transforms a generic letter into something that
could only come from you.
```
"Adjust the tone for a sudden, tragic death rather
than an expected passing."
```
Different types of loss require different pastoral responses.
```
"Shorten this to 100 words for a sympathy card
instead of a full letter."
```
You can condense longer content while preserving the care and
compassion.

🤝 **Your Discernment Matters**: AI can draft structure and suggest
language, but it doesn't know the deceased, the family's faith journey, or
the specific circumstances of the death. Always personalize with specific
memories, relationships, and pastoral knowledge. Never send an AI-
generated condolence letter without **significant** personal additions.

🔒 **Privacy Reminder**: Don't input identifying details about the
deceased or family into AI when drafting condolence letters. Use general
descriptions only.

How to Use It
1. Use AI's draft as a structural starting point
2. Add specific details about the deceased (their gifts, their impact,
   your memories)
3. Reference your actual relationship with the family

Reflection Question
What do you know about this person and this family that AI could never
know, and how can you weave that knowledge into the letter so it feels
like it could only come from you?

CHAPTER 48
Drafting Conflict Resolution Talking Points

Our Goal
Prepare clear, pastoral talking points for navigating a church conflict, helping you stay focused on facts, grace, and resolution rather than getting swept into emotional reactivity.

The Problem
Church conflict is emotionally exhausting and relationally complex. When you're in the middle of a dispute – whether it's between staff members, among church leaders, or involving congregational factions – it's hard to think clearly, but you need a plan **now**.

Why This Matters
How you navigate conflict shapes your church's culture and your credibility as a leader. Well-prepared talking points help you stay grounded in truth and grace, avoid saying something you'll regret, and guide conversations toward resolution instead of escalation. AI can't resolve conflict for you, but it can help you organize your thoughts before walking into a hard conversation.

The Prompt
```
I'm a [denomination] pastor preparing for a difficult
conversation about [brief, factual description of
conflict: e.g., disagreement over worship style,
staff role confusion, budget priorities].

The people involved are: [roles, not names: e.g.,
worship leader and long-time church member, two board
members, staff member and volunteer]

Help me prepare talking points that:
- Clearly state the issue without assigning blame
- Acknowledge each party's perspective
- Identify common ground or shared values
- Suggest 2-3 possible paths forward
- Set healthy boundaries for the conversation

Tone: Calm, fair, pastoral
Theological grounding: [Your tradition's approach to
reconciliation and community]
Length: 200-300 words of talking points
```

What to Expect
AI will generate neutral, structured talking points that help you frame the conversation. It will identify potential common ground and suggest resolution paths. The tone will be fair but may lack the specific pastoral nuance your situation requires.

Follow-Up Prompts
"Add a section on how to respond if one party becomes defensive or accusatory."
This helps you prepare for likely scenarios so you're not caught off-guard in the moment.
"Reframe the 'paths forward' section to emphasize restorative justice rather than compromise."
You can align the resolution approach with your theological convictions about reconciliation.
"Suggest an opening prayer I could use to set the tone for this conversation."
Starting with prayer grounds the conflict in spiritual practice and reminds everyone of their shared faith.

🙂 **Your Discernment Matters**: Human relationships are complex, power dynamics matter, and some conflicts require outside mediation or professional intervention. Use AI to organize your thoughts, but trust your pastoral instincts and training when you're in the room.

🔒 **Privacy Reminder**: Don't input real names or identifying details when drafting conflict resolution talking points. Use role descriptions only.

How to Use It
1. Review talking points for fairness – does each party feel heard?
2. Add specific examples or concerns you're aware of but didn't include in the prompt
3. Pray over the talking points asking God to guide the conversation

Reflection Question
What power dynamics exist in this conflict that AI couldn't account for (age, tenure, race, gender, formal or informal authority)? How will you address those dynamics pastorally during the conversation?

CHAPTER 49
Creating Crisis Response Communication Plans

Our Goal
Develop a communication plan for responding to congregational or community crises (death, tragedy, natural disaster, violence) that keeps people informed and pastorally cared for.

The Problem
When crisis strikes, people need accurate information and pastoral reassurance quickly. But in the chaos of a tragedy, you're trying to care for those directly affected while also communicating with the broader congregation.

Why This Matters
How you communicate during crisis either calms or amplifies congregational anxiety. A clear plan ensures accurate information reaches people quickly, pastoral care is coordinated, and misinformation is minimized. Creating crisis communication templates in advance, helps you respond pastorally without scrambling to figure out what to say.

The Prompt
```
I'm creating a crisis communication plan for
responding to [type of crisis: sudden death of
member, community tragedy, natural disaster, church-
related incident].

Create a communication plan that includes:
- Initial notification message (email/text) informing
congregation of the crisis
- Pastoral care message offering comfort and prayer
- Practical information about [memorial service,
emergency response, safety protocols, evacuation]
- How people can help or support those affected
- Closing with assurance of ongoing pastoral presence

Tone: Calm, compassionate, clear
Length: 200-300 words total across messages
Audience: Congregation members of all ages
```

What to Expect
AI will generate crisis communication templates with appropriate tone, structure, and pastoral care language. You'll need to customize with

actual crisis details, but the framework helps you communicate clearly when you're emotionally overwhelmed.

Follow-Up Prompts
```
"Add specific logistical details about where people
can gather for prayer and mutual support."
```
Concrete information about when and where helps people know how to respond.
```
"Create a follow-up message for 48 hours after the
initial crisis to update people on pastoral care
efforts."
```
Ongoing communication prevents people from feeling abandoned after initial shock wears off.
```
"Adjust the tone to be more urgent and directive –
this is an active safety threat."
```
Different crises require different communication urgency and tone.

🔒 **Privacy Reminder**: Share only information families have explicitly approved for congregational communication.

⚠ **Verify This**: In fast-moving crises, verify all facts before communicating. Misinformation spreads quickly and damages trust.

🫤 **Your Discernment Matters**: Crisis communication requires balancing transparency with pastoral sensitivity. AI can help you draft messages, but only you know what your congregation needs to hear.

How to Use It
1. Use AI to create crisis communication templates BEFORE crises occur
2. When a crisis hits, customize the template with specific details and send quickly
3. Ask AI to help you draft follow-up messages as the situation evolves; provide links to local updates

Reflection Question
What information does your congregation genuinely need to know right now versus what can wait until you have more facts and pastoral clarity?

CHAPTER 50
Developing Mental Health Resource Guides

Our Goal
Create accessible mental health resource guides that help congregation members find professional support, reduce stigma, and understand the church's role in mental health care.

The Problem
Mental health struggles are prevalent in every congregation, but many people don't know where to find help. You want to provide resources without overstepping your pastoral role into clinical territory. Creating a comprehensive mental health guide requires research into local resources, insurance navigation, and crisis hotlines – which is not your area of expertise and thus not a valuable use of your time.

Why This Matters
The church can be a healing community for those struggling with mental health, but only if we reduce stigma and provide clear pathways to professional care. A well-designed resource guide empowers people to seek help, educates the congregation about mental health, and clarifies that pastoral care complements (not replaces) professional treatment.

The Prompt
```
I'm creating a mental health resource guide for our
congregation in [location: city, state].

Create a guide (300-400 words) that includes:
- Brief introduction reducing stigma and encouraging
help-seeking
- Categories of resources: crisis hotlines, local
therapists/counselors, psychiatrists, support groups,
sliding-scale clinics
- How to navigate insurance and find affordable care
- When to seek emergency help vs. ongoing therapy
- The church's role: pastoral support complements
professional care
- Encouragement and hope

Tone: Compassionate, non-clinical, destigmatizing
Audience: Congregation members and families
Note: Include general categories, not specific
provider names (those change frequently)
```

What to Expect
AI will generate a structured resource guide with categories, explanations, and encouragement. You'll need to research and add specific local providers, phone numbers, and websites for your area.

Follow-Up Prompts
```
"Add a section specifically addressing resources for
parents seeking help for children or teens."
```
Different populations need different entry points to mental health care.
```
"Include guidance on how to support a loved one who's
struggling but resistant to seeking help."
```
Many people want to help but don't know how without being intrusive.
```
"Create a companion piece addressing common myths
about mental health and faith."
```
Theological misconceptions often prevent people from seeking the care they need.

🙂 **Your Discernment Matters**: You are a pastor, not a mental health professional. The resource guide should direct people to licensed therapists and psychiatrists, not position you as a counselor.

⚠ **Verify This**: Research local resources carefully. Outdated phone numbers or closed clinics undermine the guide's usefulness. Update the guide annually.

🔒 **Privacy Reminder**: If people ask you for mental health resources, protect their privacy. Don't share who's seeking help or assume you can tell others about their struggles.

How to Use It
1. Use AI's framework and research specific local resources to populate each category
2. Ask AI to create versions for different formats (printed handout, website page, email series)
3. Have AI develop companion materials (sermon series on mental health, small group discussion guides)

Reflection Question
What theological messages about mental health does your congregation need to hear to reduce stigma and encourage help-seeking?

CHAPTER 51
Writing Pre-Marital Counseling Discussion Guides

Our Goal
Create discussion guides for pre-marital counseling sessions that help couples explore important topics, build communication skills, and prepare for marriage realistically.

The Problem
Pre-marital counseling is one of the most important pastoral services you provide, but developing comprehensive discussion guides for each session takes significant time. You want to cover essential topics (communication, conflict, finances, faith, intimacy, family) without just lecturing couples or using generic workbooks that don't fit *your* theological perspective or *the couple's* specific needs.

Why This Matters
Strong marriages are built on honest conversation about hard topics. Pre-marital counseling creates space for couples to discuss what they've been avoiding, align expectations, and build skills for navigating conflict. AI can help you offer holistic and personalized counseling.

The Prompt
```
I'm creating a pre-marital counseling discussion
guide focused on [topic: communication, conflict
resolution, finances, faith practices, intimacy,
family of origin, roles and expectations].

Create a discussion guide that includes:
- Opening question to ease into the topic
- 5-7 discussion questions moving from surface to
depth
- At least 2 questions that invite vulnerability or
difference
- Reflection prompts for the couple to discuss at
home
- Brief pastoral guidance on common challenges in
this area

Theological perspective: [Your tradition's teaching
on marriage]
Tone: Warm, direct, invitational (not preachy)
Length: 250-300 words
```

What to Expect
AI will generate a complete session guide with progressive questions and pastoral insights. The questions will be solid conversation starters but may need adjustment for the specific couple's dynamics and concerns.

Follow-Up Prompts
```
"Add questions that specifically address blended
family dynamics - one partner has children from a
previous marriage."
```
Different couples need different questions based on their unique situations.
```
"Make the financial questions more specific about
debt, spending habits, and financial goal alignment."
```
Generic money questions miss the specifics that cause actual marital conflict.
```
"Include a Scripture passage and reflection questions
connecting this topic to faith."
```
Theologically grounded counseling roots marriage preparation in more than communication skills.

Your Discernment Matters: Pre-marital counseling isn't just working through a checklist. Stay attentive to red flags (controlling behavior, unresolved trauma, unrealistic expectations) that may require deeper conversation or referral to a professional counselor.

Privacy Reminder: What couples share in pre-marital counseling is confidential. Don't share their struggles or concerns with others, even church leadership, without explicit permission.

How to Use It
1. Use AI's discussion guides as your session framework and adjust questions based on each couple's specific situation
2. Ask AI to create guides for topics specific to the couple (interfaith marriage, significant age gap, cultural differences)
3. Have AI generate follow-up resources or homework assignments between sessions

Reflection Question
What conversation is this couple avoiding that needs to happen before they say "I do," and how can you create space for that discussion?

CHAPTER 52
Drafting Letters of Recommendation for Members

Our Goal
Write thoughtful letters of recommendation for church members applying for jobs, schools, volunteer positions, or professional opportunities that highlight their character and contributions.

The Problem
Church members may ask you to write recommendation letters because you've seen their character, service, and faith in action. You want to honor them with a strong letter, but writing personalized recommendations for multiple people takes time, and generic letters don't help anyone.

Why This Matters
A pastor's recommendation carries weight because you know people in contexts most don't see – their generosity, integrity, service, and how they treat others. When AI helps you draft recommendations, you can honor people's requests without sacrificing quality or spending hours per letter.

The Prompt
```
I'm writing a letter of recommendation for [name], a
member of our congregation, who is applying for
[opportunity: job, graduate school, volunteer
position, scholarship].

Write a recommendation letter (250-300 words) that:
- Introduces my relationship with the person (pastor,
how long I've known them)
- Highlights their character, integrity, and
contributions to church community
- Highlights specific examples of their service,
leadership, or growth [provide these examples]
- Connects their qualities to the opportunity they're
seeking
- Offers strong endorsement

Person's involvement: [roles they've held, ministries
they've served, notable contributions]
Opportunity type: [professional, academic, service]
Tone: Professional, warm, genuinely endorsing
```

What to Expect
AI will generate a professionally structured recommendation letter with appropriate tone and content. It will feel generic until you add specific stories and examples from your actual experience with the person.

Follow-Up Prompts
```
"Add this specific story about how this person
demonstrated servant leadership during our mission
trip."
```
Concrete examples make recommendations compelling rather than just listing generic virtues.
```
"Emphasize their resilience and growth through
personal challenges while maintaining church
involvement."
```
Character development stories show depth beyond surface-level accomplishments.
```
"Adjust the tone to be more formal and academic –
this is for a graduate school application."
```
Different opportunities require different professional registers.

👤 **Your Discernment Matters**: Only write recommendations for people you genuinely know and can honestly endorse. Declining a recommendation request is better than writing a lukewarm letter.

⚠ **Verify This**: Confirm the person's actual roles and contributions at your church before writing. Inaccurate information damages both their application and your credibility.

🔒 **Privacy Reminder**: Ask the person what they want you to emphasize or avoid mentioning. Respect their privacy about their personal life.

How to Use It
1. Use AI's letter as your structural foundation and add specific stories only you know
2. Ask AI to adjust emphasis based on what the opportunity values most (leadership, scholarship, service)
3. Have AI create recommendation letter templates for common requests (college, jobs, mission trips)

Reflection Question
What have you witnessed about this person's character in moments when they thought no one was watching?

CHAPTER 53
Creating Pastoral Care Visit Logs and Follow-Up Systems

Our Goal
Develop a simple system for logging pastoral visits and tracking follow-up care so no one falls through the cracks.

The Problem
You visit people in hospitals, homes, and care facilities regularly, but remembering who needs follow-up calls, prayer, or continued support is nearly impossible without a system. You don't want to forget people or lose track of ongoing pastoral care needs, but creating and maintaining tracking systems feels like adding administrative burden to already overwhelming pastoral work.

Why This Matters
Pastoral care doesn't end with one visit. People need ongoing prayer, follow-up calls, and continued presence through recovery, grief, or crisis. A simple tracking system ensures no one is forgotten and helps you coordinate care across your congregation. When AI helps you design this system, you can create a custom system **that works for you** and maintains consistency without complex software or time-consuming record-keeping.

The Prompt
```
I'm creating a simple pastoral care tracking system
to log visits and follow-up needs.

Design a system that includes:
- Visit log template (date, person visited, reason,
key concerns, prayer requests)
- Follow-up action items (calls to make, meals to
coordinate, future visits)
- Priority levels (urgent, routine, as-needed)
- Coordination with lay care teams or deacons
- Privacy considerations and confidentiality

Format: Simple spreadsheet or document template I can
maintain weekly [tell it what software and apps you
personally prefer]
Tone: Practical, pastor-friendly (not clinical or
institutional)
```

What to Expect
AI will generate a straightforward tracking template with fields for essential information. The system will be simple enough to maintain without becoming burdensome administrative work.

Follow-Up Prompts
"Add a section for tracking long-term ongoing care (chronic illness, grief journey, rehabilitation)."
Some pastoral care needs last months or years and require different tracking than one-time visits.
"Create a weekly review prompt that helps me identify who needs follow-up calls this week."
Regular review prevents people from being forgotten in busy weeks.
"Suggest how to coordinate this with lay care teams without sharing confidential details."
Delegating care coordination while protecting privacy requires intentional systems.

🔒 **Privacy Reminder**: Pastoral care logs contain highly sensitive information. Store them securely, don't share them carelessly, and never input identifiable details into AI when designing your system.

🧠 **Your Discernment Matters**: A tracking system helps you remember, but it can't replace pastoral discernment about who needs what kind of care when. Use the system as a tool, not a rigid protocol.

How to Use It
1. Use AI's template and customize categories based on your actual pastoral care patterns
2. Ask AI to create simplified versions if the initial system feels too complex
3. Have AI generate reminder prompts or review questions that help you stay consistent

Reflection Question
Who in your congregation might be quietly struggling right now without anyone noticing, and how could a better tracking system help you stay connected?

CHAPTER 54
Writing Prayers for Difficult Pastoral Moments

Our Goal
Craft pastoral prayers that hold grief, lament, and hope together during tragic moments when easy words fail.

The Problem
When tragedy strikes – sudden death, violence, injustice, community crisis – people look to you for words that make sense of the senseless. But there are no easy answers, and platitudes feel impersonal. You need prayers that acknowledge the full weight of suffering while still affirming God's presence. Finding those words when you're also grieving or overwhelmed can be nearly impossible.

Why This Matters
Pastoral prayers during tragedy give voice to what people feel but can't articulate. They create space for honest lament, validate suffering, and gently hold the tension between despair and hope. When AI helps you draft these prayers, you can find language for the unspeakable without rushing to false comfort.

The Prompt
```
I'm writing a pastoral prayer for [tragic situation:
sudden death, community violence, natural disaster,
injustice, collective grief].

Write a prayer (200-250 words) that:
- Honestly names the tragedy and pain
- Gives voice to lament and questions
- Acknowledges God's presence without offering easy
answers
- Holds grief and hope in tension
- Ends with commitment to one another and trust in
God

Theological perspective: [Your tradition's approach
to suffering and theodicy]
Tone: Honest, tender, not falsely comforting
Avoid: Platitudes, "everything happens for a reason,"
minimizing pain
```

What to Expect
AI will generate a prayer that balances lament with hope, honesty about suffering with trust in God. The prayer will need personalization with specific details about the actual tragedy and community.

Follow-Up Prompts
"Make this prayer more specific to the grief our congregation feels about this particular loss."
Generic tragedy prayers miss the specific horror and heartbreak of what actually happened.
"Add language that acknowledges anger at God – people need permission to bring their rage to prayer."
Honest prayers create space for the full range of human emotion in God's presence.
"Include a section naming our commitment to work for justice in response to this tragedy."
Lament can lead to action, not just passive acceptance of suffering.

Your Discernment Matters: Prayers during tragedy require extreme pastoral sensitivity. Don't rush to hope or resurrection before the community has had space to lament and grieve.

Theological Caution: Avoid theological explanations that blame victims, minimize suffering, or suggest God caused the tragedy "for a reason." Mystery and lament are more faithful than false certainty.

Privacy Reminder: When praying about specific tragedies involving individuals, respect their privacy and dignity. Don't share details families haven't made public.

How to Use It
1. Use AI's prayer as your starting framework and make it specific to your community's actual grief
2. Ask AI to adjust theological emphasis based on your tradition's approach to suffering and/or evil
3. Have AI create follow-up prayers for ongoing grief (one week later, one month later)

Reflection Question
What does your congregation most need to hear from God in this moment – comfort, permission to grieve, assurance of presence, call to action, or something else?

CHAPTER 55
Developing Grief Support Group Curriculum

Our Goal
Create a multi-week grief support group curriculum that guides participants through shared grief while honoring each person's unique journey.

The Problem
Grief support groups provide essential community for those mourning, but developing curriculum that balances structure with flexibility, education with space for tears, and hope with honesty is complex work. You want to offer more than just a gathering place – you want intentional support that helps people process loss – but creating 3-8 weeks of content is a **lot** of work.

Why This Matters
Support groups remind mourners they're not alone and give them tools for navigating loss. Well-designed curriculum provides just enough structure to feel safe while leaving space for individual stories and emotions. When AI helps you develop this curriculum, you can offer substantive grief support without spending months on preparation.

The Prompt
```
I'm creating a [3-week/8-week] grief support group
curriculum for people mourning [type of loss: death
of spouse, child, parent, or general grief group].

Create a curriculum outline that includes:
Session themes (e.g., acknowledging loss, navigating
holidays, guilt and regrets, finding meaning, hope
and moving forward)
Discussion questions for each session
Brief educational content about grief stages/process
Activities or reflections participants can do between
sessions
Scripture or spiritual resources when appropriate

Theological tradition: [Your approach to grief,
resurrection, and hope]
Group context: [size, mix of recent/long-term grief,
ages]
Tone: Compassionate, honest, space-holding (not
fixing or rushing)
```

What to Expect
AI will generate a complete curriculum outline with session themes, discussion questions, and activities. You'll need to personalize based on your group's actual composition and grief experiences.

Follow-Up Prompts
"Add specific guidance for Session 3 on how to handle the holidays and anniversary dates."
Practical coping strategies help people navigate grief triggers in daily life.
"Include questions that create space for anger, guilt, and complicated grief – not just sadness."
Grief involves many emotions beyond sadness, and curriculum should honor that complexity.
"Suggest how to adapt Session 6 for people whose grief is disenfranchised (miscarriage, pet loss, estranged relationships)."
Some grief isn't socially validated; those mourners need particular care.

🤔 **Your Discernment Matters**: Grief support groups aren't therapy. Know when someone's grief requires professional counseling beyond what a support group can provide.

⚠ **Verify This**: Educate yourself about complicated grief, trauma, and when to refer people for professional help. Don't rely solely on AI's general grief information.

🔒 **Privacy Reminder**: What's shared in grief groups is confidential. Establish clear group agreements about privacy and honor them rigorously.

How to Use It
1. Use AI's curriculum outline and adjust session themes based on your group's actual needs
2. Ask AI to create supplemental resources (reading lists, journaling prompts, prayer guides)
3. Have AI develop follow-up support materials for after the group ends

Reflection Question
What would help this group feel safe enough to share their deepest grief, and how can your curriculum create that space?

CHAPTER 56
Adapting Sermon Messages to Address a Community Crisis

Our Goal
Quickly adapt your planned sermon to address an unexpected community crisis while still maintaining Biblical faithfulness and pastoral sensitivity.

The Problem
Your sermon is ready, but between Saturday night and Sunday morning, crisis strikes – a community tragedy, national disaster, local violence, or shocking news. Your planned sermon suddenly feels disconnected from what people are actually experiencing. You need to address the crisis pastorally, but you don't have time to write an entirely new sermon.

Why This Matters
When crisis happens, people come to church asking "Where is God in this?" Preaching as if nothing happened feels tone-deaf and pastorally negligent. But abandoning your sermon entirely creates last-minute panic. When AI helps you adapt your message, you can address the crisis faithfully without starting from scratch Saturday night.

The Prompt
```
I planned to preach on [Scripture passage] about
[theme], but [crisis event] just happened in our
community.

Help me adapt my sermon to address this crisis by:
- Suggesting how my planned Scripture connects to
this moment
- Offering a revised introduction that acknowledges
the crisis
- Providing pastoral framing that holds both the
planned message and immediate need
- Suggesting what to cut or adjust from my original
sermon
- Proposing a closing that speaks to current
grief/fear/uncertainty

Original sermon focus: [brief summary]
Crisis details: [general description, not
confidential specifics]
Congregation's likely emotional state: [shock, grief,
fear, anger]
Length: Keep the same time frame as original sermon
```

What to Expect
AI will suggest how to bridge your planned message with the immediate crisis, offering revised introductions, connections, and applications. You'll maintain Biblical faithfulness while addressing pastoral urgency.

Follow-Up Prompts
"Add specific language acknowledging that we don't have answers yet and it's okay to sit with mystery."
Crisis sermons often need to give permission for uncertainty rather than rushing to theological conclusions.
"Suggest how to move from lament in the sermon to concrete ways the congregation can respond or help."
Grief can lead to action, and sermons can point toward meaningful response.
"Adjust the tone to be more directive and reassuring - people need clear guidance about safety or next steps."
Some crises require pastoral authority and clear direction, not just comfort.

🤔 **Your Discernment Matters**: Not every crisis requires sermon revision. Sometimes the planned message is exactly what people need. Trust your pastoral instinct about when to adapt.

⚠ **Verify This**: In fast-moving crises, verify facts before preaching about them. Misinformation from the pulpit damages credibility and spreads confusion, gossip, or even panic.

📖 **Scripture Check**: Ensure the Biblical text actually speaks to the crisis in faithful ways. Don't force connections that aren't exegetically sound.

How to Use It
1. Ask AI to help you identify which parts of your planned sermon to cut for time
2. Have AI create follow-up pastoral care resources addressing the crisis for the week ahead
3. Have AI create one or more relevant social media posts for you

Reflection Question
What does your congregation most need to hear from God in this crisis moment, and how can your planned Scripture reading speak to that need?

CHAPTER 57
Drafting Pastoral Letters to Address Sensitive Topics

Our Goal
Write pastoral letters that address sensitive congregational or community topics with honesty, grace, and theological grounding when issues are too complex for a Sunday announcement.

The Problem
Some topics require more space than Sunday morning allows – pastoral transitions, theological controversies, denominational conflicts, congregational disputes, or addressing harm. You need to communicate clearly about difficult subjects while maintaining pastoral tone, theological depth, and care for diverse perspectives. Writing these letters can take time you might not have, yet they're too important to rush.

Why This Matters
Pastoral letters set the tone for how congregations navigate difficulty. Well-written letters educate, comfort, challenge, and invite unity while demonstrating leadership that honors people's intelligence and faith. AI can help you communicate thoughtfully about sensitive topics without sacrificing pastoral care or theological nuance.

The Prompt
```
I'm writing a pastoral letter to the congregation
about a sensitive topic [pastoral transition,
theological controversy, denominational debate,
congregational conflict, addressing harm, etc.].

Write a pastoral letter (400-500 words) that:
- Opens with pastoral warmth and acknowledgment of
difficulty
- Clearly explains the situation with honesty
- Provides theological/Biblical grounding for the
church's response
- Acknowledges diverse perspectives and emotions
- Offers clear next steps or pastoral care resources
- Closes with hope and invitation to continued
conversation

Theological tradition: [Your denomination and
perspective]
Congregation context: [size, diversity, history with
this issue]
```

```
Tone: Pastoral, honest, unifying (not defensive or
dismissive)
```

What to Expect
AI will generate a complete pastoral letter with appropriate structure, tone, and theological grounding. You'll need to personalize with specific congregational details and your authentic pastoral voice.

Follow-Up Prompts
```
"Add a section acknowledging the pain this situation
has caused specific groups in our congregation."
```
Naming harm directly demonstrates pastoral care and creates space for healing.
```
"Adjust the tone to be more direct and less diplomatic
- this requires clear leadership, not just conversation."
```
Some situations require pastoral authority and clear direction, not just dialogue.
```
"Include specific Scripture passages and brief
reflections grounding our response Biblically."
```
Theological grounding helps people see the issue through a faith lens, not just institutional logistics.

🗨️ **Your Discernment Matters**: Pastoral letters are public documents that can be shared beyond your congregation. Write with awareness that your words may reach unintended audiences.

✝️ **Theological Caution**: Ensure your letter accurately represents your denomination's positions while remaining sensitive to those who disagree.

🔒 **Privacy Reminder**: Don't share identifying details about individuals involved in sensitive situations without their explicit permission.

How to Use It
1. Use AI's letter as your foundation and infuse it with your specific pastoral voice and congregational knowledge
2. Ask AI to create versions emphasizing different aspects (unity, truth-telling, pastoral care) and choose the best balance
3. Have AI draft follow-up communication if the situation continues to develop

Reflection Question
What truth needs to be told in this letter that people might not want to hear, and how can you say it with both honesty and grace?

SECTION 5

EDUCATION & DISCIPLESHIP

CHAPTER 58
Creating Small Group Discussion Questions from Sermons

Our Goal
Transform your Sunday sermon into discussion questions that help small groups dive deeper into Scripture and apply it to their lives.

The Problem
You've spent hours preparing a sermon, but creating quality small group materials feels like starting from scratch. By the time Sunday is over, you're mentally exhausted and the idea of writing discussion questions feels overwhelming.

Why This Matters
Small groups are where transformation happens. When you provide discussion guides connected to your preaching, you create continuity between Sunday worship and weekday discipleship. Repurposing your sermon into small group content multiplies its impact without multiplying your workload.

The Prompt
```
I preached a sermon on [Scripture passage] with the
following main points:
1. [Main point 1]
2. [Main point 2]
3. [Main point 3]

[If possible, attach a PDF of your sermon script.]

Create a small group discussion guide for [audience:
e.g., adult small groups, youth groups, young adults]
that includes:
- An opening icebreaker question (light and
relational)
- 4-6 discussion questions that move from observation
→ interpretation → application
- At least 2 questions that ask for personal stories
or examples
- A closing question that invites action or
commitment
- A brief prayer prompt related to the theme

Tone: Conversational, inviting, thought-provoking
```

```
Avoid: Overly academic language, yes/no questions,
questions with obvious answers
```

What to Expect

AI will generate 6-8 questions with good progression from "What does this passage say?" to "How will you live differently this week?" The icebreaker will be generic but functional.

Follow-Up Prompts

```
"Make question #3 more personal and vulnerable."
```
This helps you adjust the tone to encourage deeper, more honest sharing in your groups.
```
"Create a version of this guide for teenagers instead
of adults."
```
You can take the same Biblical content and quickly adapt it for different age groups or spiritual maturity levels.
```
"Add a question that connects this Scripture to
current events or cultural issues."
```
AI can help you make ancient texts feel relevant to what your congregation is experiencing right now.

📖 **Scripture Check**: Ensure discussion questions actually reflect the passage's meaning in context. AI sometimes creates questions that sound spiritual but don't align with the text's original intent.

🧠 **Your Discernment Matters**: Generic questions won't create the kind of vulnerable sharing that builds community. Personalize with examples from your congregation's actual life together.

How to Use It

1. Use AI's questions as your foundation and adjust based on your group's spiritual maturity
2. Ask AI to create multiple versions and pick the best questions from each
3. Have AI generate follow-up questions for groups that want to go deeper

Reflection Question

Which question would be most challenging for your congregation to answer honestly, and how might you reframe it to create safety for vulnerable sharing?

CHAPTER 59
Developing Bible Study Plans (4-Week, 8-Week, 90-Day)

Our Goal
Create structured Bible study plans that guide individuals or groups through Scripture with clear daily readings, reflection questions, and application prompts.

The Problem
People want to read the Bible more consistently but don't know where to start or how to stay on track. You want to provide a reading plan that's accessible, sustainable, and formative, but designing multi-week study guides with daily content takes significant time.

Why This Matters
Bible study plans give people structure for spiritual formation. A well-designed plan provides just enough guidance to stay engaged without becoming overwhelming. When AI helps you create these plans, you can offer robust Biblical discipleship without spending weeks on curriculum development.

The Prompt
```
I'm creating a [4-week/8-week/90-day] Bible study
plan focused on [Scripture book, theme, or topic:
e.g., Gospel of Mark, Psalms of lament, prophetic
justice, spiritual practices].

Create a study plan that includes:
- Daily Scripture readings (manageable portions)
- Weekly themes that build progressively
- 2-3 reflection questions per reading
- Brief context or background information when needed
- Weekly application challenges or spiritual
practices

Target audience: [individuals, small groups, new
believers, mature disciples]
Reading level: [accessible to spiritual seekers,
appropriate for theological depth]
Format: [daily readings, weekly gathering discussion,
both]
```

What to Expect
AI will generate a complete study plan with daily readings, themes, questions, and application prompts. The plan will be well-structured but generic until you add your theological perspective and congregational context.

Follow-Up Prompts
```
"Add specific spiritual practices for each week
(prayer, fasting, service, contemplation) that
connect to the weekly theme."
```
Integrating practices beyond just reading helps people embody what they're learning.
```
"Create discussion questions for weekly small group
gatherings based on the week's readings."
```
If groups are studying together, they need different questions than individuals need for personal reflection.
```
"Adjust Week 3 to address questions or doubts people
commonly have about this Scripture."
```
Good study plans anticipate where people will struggle and provide help.

📖 **Scripture Check**: Verify that daily readings and reflection questions accurately represent the Biblical text in context, not just proof-texting themes.

🤔 **Your Discernment Matters**: A reading plan that's too ambitious (too long, too academic, too demanding) sets people up for failure. Keep it manageable and enjoyable for *your* audience.

✝️ **Theological Caution**: Ensure your study plan reflects sound Biblical interpretation and your theological tradition's approach to Scripture.

How to Use It
1. Use AI's study plan framework and adjust reading portions based on what's realistic for your congregation
2. Ask AI to create companion resources (journaling prompts, prayer guides, reading schedules)
3. Have AI develop leader guides if groups are studying together

Reflection Question
What pace of Bible reading would actually be sustainable for your congregation, and does this plan honor that reality or set unrealistic expectations?

CHAPTER 60
Writing Curriculum for Confirmation Classes

Our Goal

Create confirmation curriculum that teaches core Christian beliefs, denominational distinctives, and spiritual practices in age-appropriate, engaging ways.

The Problem

Confirmation is one of the most important teaching opportunities you have with young people, but developing 8-12 weeks of curriculum that's theologically sound, developmentally appropriate, and actually engaging is massive work. You want confirmands to understand faith deeply, not just memorize answers.

Why This Matters

Confirmation shapes young people's understanding of faith, church, and their place in God's creation and family. Well-designed curriculum helps them claim faith as their own, not just their parents'. When AI helps you create confirmation materials, you can offer substantive theological education without spending months on lesson plans.

The Prompt

```
I'm creating confirmation curriculum for [age group:
middle school, high school] in the [denomination]
tradition.

Create a [8-week/12-week] curriculum outline that
includes:
- Weekly session themes (Trinity, salvation,
sacraments, church, mission, discipleship, etc.)
- Core questions each session addresses
- Scripture passages and theological concepts
- Discussion questions and activities
- Application to teenagers' actual lives
- Denominational distinctives where relevant

Theological tradition: [Your denomination's
confirmation theology]
Session length: [60 minutes, 90 minutes]
Teaching approach: [discussion-based, interactive,
includes service projects]
```

What to Expect
AI will generate a complete curriculum outline with weekly themes, discussion questions, activities, and Scripture. You'll need to personalize with your denomination's specific practices and your group's maturity level.

Follow-Up Prompts
```
"Add specific activities for Session 4 on the
sacraments that let students experience baptism and
communion practices."
```
Experiential learning helps abstract theology become concrete and memorable.
```
"Create take-home questions for parents to discuss
with their confirmands each week."
```
Involving parents makes confirmation a family faith journey, not just a church program.
```
"Adjust Session 7 to address doubts and questions
teenagers actually have about faith."
```
Good confirmation doesn't just teach answers – it creates space for honest questions.

📖 **Scripture Check**: Confirmation curriculum must be Biblically grounded and theologically sound. Ensure all Scriptures fit well.

✝️ **Theological Caution**: Ensure curriculum accurately represents your denomination's beliefs and practices, especially on contested topics like salvation or sacraments.

🙂 **Your Discernment Matters**: Not all confirmands are at the same place spiritually or intellectually. Meet individuals where they are.

How to Use It
1. Use AI's curriculum outline and adapt for your denomination's specific confirmation requirements
2. Ask AI to create supplemental materials (parent guides, Scripture memory cards, service project ideas)
3. Have AI develop assessment tools that honor faith development, not just knowledge acquisition

Reflection Question
What do you most want these young people to understand about faith and church by the time they're confirmed?

CHAPTER 61
Creating New Member Orientation Materials

Our Goal
Develop new member orientation materials that welcome people into your church's life, explain your values and practices, and help them find their place in the community.

The Problem
New members join with excitement but often don't know how to get connected, what your church believes, how decisions are made, or where they can serve. You want to provide thorough orientation without overwhelming people or making it feel like institutional onboarding.

Why This Matters
The first 90 days often determine whether new members stick or drift away. Well-designed orientation materials answer common questions, reduce barriers to involvement, and help people find belonging. AI can help you offer consistent, quality onboarding for every new member.

The Prompt
```
I'm creating new member orientation materials for our
[denomination] church.

Create orientation content that includes:
- Welcome letter from the pastor
- Brief church history and mission
- Core beliefs and values
- Governance structure and decision-making
- Ministry opportunities and how to get involved
- Small groups, worship practices, and discipleship
pathways
- Practical information (giving, building access,
communication)
- Next steps for deeper connection

Church context: [size, denominational tradition,
ministry focus]
Format: [printed handbook, email series, orientation
class sessions]
Tone: Welcoming, informative, invitational (not
institutional)
Length: 300-400 words total across sections
```

What to Expect

AI will generate comprehensive orientation content with appropriate structure and tone. The materials will be generic until you add specific details about your actual church, leadership, and ministries.

Follow-Up Prompts

```
"Add a section specifically for people transferring
from other denominations explaining our distinctive
practices."
```

Transfer members need different orientation than those new to church altogether.

```
"Create a new member FAQ addressing common questions
about membership expectations."
```

Proactively answering questions reduces anxiety and clarifies expectations.

```
"Develop a 4-week email series delivering orientation
content incrementally rather than all at once."
```

Spreading information over time prevents overwhelm and increases retention.

✝ **Theological Caution**: Ensure your beliefs statement accurately represents your denomination's theology and your congregation's actual convictions.

⚠ **Verify This**: Keep contact information, ministry details, and leadership names current. Outdated orientation materials create confusion.

🙂 **Your Discernment Matters**: New members come from diverse backgrounds and need different levels of explanation. Avoid assuming insider knowledge.

How to Use It

1. Use AI's orientation content and customize with your church's specific history, values, and opportunities
2. Ask AI to create versions for different formats (handbook, website, presentation slides)
3. Have AI develop follow-up materials for 30, 60, and 90 days post-membership

Reflection Question

What do new members most need to know to feel like they belong here, and are you giving them that information clearly?

CHAPTER 62
Developing Book Discussion Guides for Church Book Clubs

Our Goal
Create discussion guides for church book clubs that foster meaningful conversation, Biblical reflection, and community building around shared reading.

The Problem
Book clubs are popular discipleship tools, but facilitating good discussions requires more than "What did you think?" You need questions that go deeper, connect books to faith, and invite vulnerable sharing. Creating quality discussion guides for every book your groups read is time-consuming.

Why This Matters
Book clubs combine intellectual engagement with community formation. Well-designed discussion guides help people connect what they're reading to their faith journey and to one another. When AI helps you create these guides, you can offer substantive discussion resources without spending hours on question development.

The Prompt
```
I'm creating a discussion guide for a church book
club reading [book title and author].

The book's main themes are: [brief summary of key
ideas or topics]

Create a discussion guide that includes:
- Opening question to ease into conversation
- 6-8 discussion questions moving from surface to
depth
- Questions connecting the book's themes to faith,
Scripture, or spiritual practice
- At least 2 questions inviting personal reflection
or story-sharing
- Closing question about application or next steps

Book club context: [demographics, spiritual maturity,
reading history]
Tone: Conversational, thought-provoking, welcoming to
diverse perspectives
```

What to Expect
AI will generate thoughtful discussion questions with good progression from observation to application. The questions will be generic until you connect them to your congregation's specific context and concerns.

Follow-Up Prompts
"Add questions specifically connecting Chapter 3's argument about suffering to our Wesleyan theology of redemption."
Denominational or theological lenses help people read through their tradition's perspective.
"Create questions that address where readers might disagree with the author's conclusions."
Good discussion guides make space for respectful disagreement and diverse perspectives.
"Suggest Scripture passages to pair with each discussion question for Biblical grounding."
Connecting secular books to Scripture helps people discern and integrate insights faithfully.

🗨 **Your Discernment Matters**: Not every popular book aligns with Christian faith. Be discerning about which books to read and how to engage critically with problematic content.

✝ **Theological Caution**: Discussion guides should help people evaluate books through a their lens of faith (remember the Quadrilateral!), not just accept all ideas uncritically.

How to Use It
1. Use AI's discussion guide and add questions specific to chapters or themes most relevant to your group
2. Ask AI to create guides for different types of books (theology, memoir, fiction, justice, spiritual formation)
3. Have AI develop facilitator notes for leading discussions well

Reflection Question
What conversation does this book need to spark in your congregation that might not happen otherwise, and are your questions opening that door?

CHAPTER 63
Writing Devotional Content for Lent or Advent

Our Goal
Create daily devotional content for Lent or Advent that guides your congregation through the season with Scripture, reflection, and spiritual practice.

The Problem
Seasonal devotionals are powerful discipleship tools, but writing 40 days of Lenten reflections or 28 days of Advent content requires sustained creative energy you may not have during already-busy seasons. You want to offer your congregation meaningful daily content without burning out.

Why This Matters
Daily devotionals meet people in their actual lives – morning coffee, evening quiet, commute time. They create shared spiritual rhythm across the congregation and deepen engagement with the season's themes. When AI helps you create devotional content, you can offer rich formation without sacrificing your own spiritual health during demanding seasons.

The Prompt
```
I'm creating a [Lent/Advent] devotional with daily
entries from [start date] through [end date].

For each day, create a devotional entry (150-200
words) that includes:
- A brief Scripture passage (vary between Gospels,
Psalms, prophets, Epistles)
- A short reflection connecting Scripture to
[season's themes: Lent - repentance, sacrifice,
preparation; Advent - hope, waiting, incarnation]
- A closing prayer or spiritual practice prompt
- [Optional: A question for personal reflection]

Theological tradition: [Your denomination]
Congregation context: [demographics, spiritual
maturity]
Tone: Reflective, honest, hopeful

Create the first week's devotionals (Days 1-7). I'll
request subsequent weeks as needed.
```

What to Expect
AI will generate devotional entries with Scripture, reflection, and prayer for each day. The reflections will be solid but generic until you personalize with local stories or current events.

Follow-Up Prompts
"Create Week 2 devotionals with more emphasis on the theme of wilderness and temptation."
You can build thematic progression across weeks while maintaining format consistency.
"Add a specific spiritual practice suggestion for each day (silence, fasting, service, gratitude)."
This moves people beyond reading to actually practicing Lenten or Advent disciplines.
"Adjust the reading level for teenagers - I want youth to use this too."
You can create age-appropriate versions of the same seasonal journey.

📖 **Scripture Check**: Verify that Scripture passages actually support the seasonal themes and devotional reflections. Don't force connections.

✝ **Theological Caution**: Lent and Advent have deep theological meaning beyond generic spiritual reflection. Ground devotionals in your tradition's understanding of these seasons.

🧑 **Your Discernment Matters**: Devotional writing requires your pastoral voice and knowledge of what your congregation is experiencing. Don't let AI content feel generic or disconnected.

How to Use It
1. Use AI to create devotionals in weekly batches rather than all at once
2. Ask AI to adjust themes or emphases based on current events or congregational needs
3. Have AI create companion discussion guides for small groups using the same devotional

Reflection Question
What spiritual practice or discipline do you personally need during this season, and how might that shape the devotionals you offer your congregation?

CHAPTER 64
Creating Sunday School Lesson Plans for Various Ages

Our Goal
Develop age-appropriate Sunday school lesson plans that teach Biblical stories, theological concepts, and spiritual practices in engaging, accessible ways.

The Problem
Sunday school teachers need quality lesson plans, but creating curriculum for multiple age groups (preschool, elementary, youth, adults) requires understanding developmental stages, learning styles, and age-appropriate activities. Producing this content weekly or monthly can be overwhelming for volunteer teachers or part-time youth pastors.

Why This Matters
Sunday school shapes faith formation across generations. Well-designed lessons help children encounter Biblical stories, youth wrestle with theological questions, and adults deepen understanding. When AI helps you create lesson plans, you can support teachers with quality curriculum that's both educationally sound and spiritually formative.

The Prompt
```
I'm creating a Sunday school lesson for [age group:
preschool (3-5), elementary (6-10), youth (11-17),
adults] based on [Scripture passage or theme].

Create a lesson plan (45-60 minutes) that includes:
- Opening activity or icebreaker
- Bible story or teaching (age-appropriate
presentation)
- Discussion questions or reflection prompts
- Activity, craft, or game reinforcing the lesson
- Application to daily life
- Closing prayer or blessing

Age group: [specify]
Teaching approach: [storytelling, discussion, hands-
on, experiential]
Space/resources: [classroom, limited supplies,
outdoor option]
```

What to Expect
AI will generate a complete lesson plan with activities, discussions, and applications appropriate for the age group. You'll need to adjust based on your actual students' maturity and your available resources.

Follow-Up Prompts
```
"Add more hands-on activities for kinesthetic
learners - sitting and listening doesn't work for
this group."
```
Different learners need different engagement strategies to stay focused and retain content.
```
"Create a take-home activity sheet parents can use to
reinforce the lesson during the week."
```
Involving families extends Sunday school learning into daily life.
```
"Adjust this lesson for a multi-age classroom
combining elementary and youth."
```
Small churches often have combined classes and need flexible lesson plans.

⌨ **Scripture Check**: Ensure Bible stories are told accurately and age-appropriately without distorting the text or adding details that aren't there.

👁 **Your Discernment Matters**: Not all AI-suggested activities are safe, practical, or appropriate for your specific group. Review carefully before implementing.

✝ **Theological Caution**: Children's lessons must be theologically sound while being developmentally appropriate. Avoid oversimplifying in ways that create theological misconceptions.

How to Use It
1. Use AI's lesson plan as your foundation and adjust activities based on your space and resources
2. Ask AI to create lesson series (4-6 weeks) with connected themes rather than standalone lessons
3. Have AI develop teacher training materials explaining how to lead each age group effectively

Reflection Question
What does this age group most need to learn about God right now, and does this lesson actually teach that or just keep them busy?

CHAPTER 65
Developing Discipleship Pathways and Next Steps

Our Goal
Create clear discipleship pathways that help people move from first-time visitors to mature disciples, with concrete next steps at each stage.

The Problem
People join your church wanting to grow spiritually, but they don't know what to do next. You have ministries and programs, but no clear pathway connecting them. People get stuck or drift away because they can't see the next step. Designing a coherent discipleship system that moves people from exploring faith to mature leadership is complex strategic work.

Why This Matters
Discipleship pathways eliminate confusion and create momentum. When people know "I'm here, and the next step is there," they're more likely to keep growing. Clear pathways also help leaders intentionally guide people rather than just managing programs. When AI helps you design these systems, you can create coherent discipleship architecture without hiring consultants.

The Prompt
```
I'm creating a discipleship pathway for our
[denomination] church that moves people from first
contact to mature discipleship.

Design a pathway system that includes:
- 4-6 stages of spiritual growth (e.g., Exploring,
Connecting, Growing, Serving, Leading)
- Clear descriptors of each stage
- Specific next steps and ministry offerings for each
stage
- How people know when to move to the next stage
- Support structures (mentoring, classes, groups) at
each level

Church context: [size, demographics, existing
ministries]
Theological framework: [Your tradition's
understanding of spiritual formation]
Format: [visual pathway map, written guide, both]
```

What to Expect

AI will generate a complete discipleship framework with stages, descriptors, and ministry connections. The system will be logically structured but generic until you map your actual ministries and programs to each stage.

Follow-Up Prompts

"Add specific ministry examples from our church for each pathway stage."

Generic stages become useful when connected to your actual small groups, classes, and service opportunities.

"Create assessment questions people can use to identify which stage they're currently in."

Self-assessment helps people locate themselves on the pathway and choose appropriate next steps.

"Suggest how to communicate this pathway to the congregation without it feeling institutional or programmatic."

Discipleship is relational, not just a system – framing matters.

🧠 **Your Discernment Matters**: Discipleship pathways are tools, not formulas. People grow at different paces and through different means. Don't let the system become rigid or prescriptive.

✝️ **Theological Caution**: Ensure your pathway reflects Biblical understanding of spiritual formation, not just institutional membership progression.

How to Use It

1. Use AI's pathway framework and map your existing ministries to appropriate stages
2. Ask AI to create communication materials (infographics, brochures, website content) explaining the pathway
3. Have AI develop training materials for staff and volunteers who guide people through stages

Reflection Question

What's the current roadblock preventing people from taking the next step in discipleship at your church, and does this pathway address it?

CHAPTER 66
Writing Sermon-Based Discussion Questions for the Sunday Worship Service Bulletin

Our Goal
Create brief, accessible discussion questions for worship bulletins that help families or individuals reflect on the sermon during the week.

The Problem
You want Sunday's sermon to continue shaping people's faith beyond the sanctuary, but most people forget the message by Monday. Including discussion questions in the bulletin gives families conversation starters and individuals reflection prompts, but writing these weekly adds another task to sermon prep.

Why This Matters
Bulletin discussion questions extend your preaching into homes, cars, and dinner tables. They help parents disciple children, couples discuss faith together, and individuals apply sermons to daily life. When AI helps you create these questions, you can offer weekly discipleship tools without extra sermon prep burden.

The Prompt
```
I'm preaching on [Scripture passage] with the main
message: [one-sentence summary].

[If possible, attach a PDF of your sermon script.]

Create 3-8 discussion questions for the worship
bulletin that:
- Are brief enough to fit in bulletin space (one line
each)
- Work for diverse audiences (families, couples,
individuals)
- Move from understanding to application
- Invite reflection on the sermon's main point
- Are accessible to children and adults

Congregation: [demographics, mix of ages]
Tone: Conversational, invitational, brief
```

What to Expect

AI will generate 3-8 concise questions suitable for bulletin printing. The questions will be generic until you make them specific to your actual sermon content and illustrations.

Follow-Up Prompts

```
"Make question #2 more specific to the story I told
about the homeless ministry." [explain the story]
```
Questions tied to your actual sermon moments feel more relevant and memorable.
```
"Create a version specifically for parents to discuss
with elementary-age children."
```
Family-focused questions need simpler language and more concrete examples.
```
"Add a closing question that invites specific action
this week, not just reflection."
```
Application questions move people from thinking to doing.

📖 **Scripture Check**: Ensure questions actually reflect the sermon's Biblical message, not just generic spiritual reflection.

🤔 **Your Discernment Matters**: Bulletin space is limited. Choose questions that matter most, not just fill space.

How to Use It

1. Use AI's questions and select the ones that best capture your sermon's heart
2. Ask AI to create questions weekly as part of your sermon prep workflow
3. Have AI generate different question sets for different audiences (families, singles, seniors, teens, small children)

Reflection Question

If someone only remembered one thing from your sermon and discussed one question at home, which question would create the most meaningful conversation?

CHAPTER 67
Creating Catechism or Theology Study Guides

Our Goal

Develop theology study guides or catechism materials that teach core Christian doctrines in clear, accessible language for new believers or confirmation students.

The Problem

New Christians and confirmands need solid theological grounding, but most catechisms feel archaic or overly academic. You want to teach theology clearly without dumbing it down or boring people. Creating accessible materials that honor both theological depth and contemporary language takes significant theological and pedagogical skill.

Why This Matters

Theological literacy shapes how people live their faith. Discussing and understanding concepts such as the Trinity, incarnation, atonement, and sanctification isn't just intellectual exercise – it forms identity and practice. When AI helps you create study guides, you can offer substantive theological education in language today's learners can actually understand.

The Prompt

```
I'm creating a theology study guide for [audience:
new believers, confirmation students, adult
education] covering [topic: Trinity, salvation,
church, sacraments, eschatology].

Create a study guide that includes:
- Core theological question being addressed
- Biblical foundations (key Scripture passages)
- Clear explanation of the doctrine (200-250 words)
- Explain denominational differences on this topic
- Common questions or misconceptions
- Application to daily Christian life
- Reflection or discussion questions

Theological tradition: [Your denomination's
perspective]
Reading level: [accessible to non-academics]
Tone: Clear, inviting, theologically sound (not
simplistic)
```

What to Expect
AI will generate theologically sound content with Scripture, explanation, and application. You'll need to verify theological accuracy and adjust for your tradition's specific emphases.

Follow-Up Prompts
```
"Add specific denominational distinctives about
baptism – our Wesleyan understanding differs from
Reformed theology."
```
Theological study guides must reflect your tradition's actual teaching, not generic Christianity.
```
"Include questions that address doubts or struggles
people commonly have with this doctrine."
```
Good theology teaching acknowledges where people genuinely struggle to believe or understand.
```
"Create a simplified version for middle school
confirmation students."
```
Different audiences need different levels of theological complexity and vocabulary.

📖 **Scripture Check**: Theological teaching must be grounded in Scripture and represent texts faithfully in context.

✝ **Theological Caution**: Verify all doctrinal content with trusted theological resources. AI can generate plausible-sounding theology that does not align with your denomination, or is actually heretical.

🙏 **Your Discernment Matters**: Theology isn't just information transfer – it's formation. Ensure guides invite people into deeper relationship with God, not just intellectual knowledge.

How to Use It
1. Use AI's theological content as a starting draft and verify accuracy with denominational resources
2. Ask AI to create study guides for a full theology series
3. Have AI develop leader guides for teaching these materials in classes

Reflection Question
What theological misunderstanding is most common in your congregation, and does this study guide address it clearly?

CHAPTER 68
Developing Youth Group Lessons and Activities

Our Goal
Create engaging youth group lessons that teach Biblical truth, build community, and meet teenagers where they actually are in faith and life.

The Problem
Youth ministry requires lessons that are Biblically sound, developmentally appropriate, and actually interesting to teenagers – a difficult balance. You need activities that engage without being gimmicky, discussions that go deep without being preachy, and applications that feel relevant to teenage life. Creating this content each week can be exhausting.

Why This Matters
Youth group is often where teenagers decide whether faith is real or just what their parents believe. Well-designed lessons create space for honest questions, authentic community, and genuine encounters with God. When AI helps you create youth content, you can offer quality ministry without burning out volunteer leaders or staff.

The Prompt
```
I'm creating a youth group lesson for [age: middle
school 11-14, high school 14-18] on [topic: identity,
relationships, suffering, doubt, mission, social
media, etc.].

Create a lesson plan (60-90 minutes) that includes:
- Opening game or icebreaker
- Discussion starter (video, current event, relatable
scenario)
- Biblical teaching on the topic (Scripture and
application)
- Small group discussion questions
- Activity or challenge applying the lesson
- Closing reflection or prayer

Youth group context: [size, spiritual diversity,
church background]
Teaching approach: [discussion-heavy, experiential,
service-focused]
Tone: Authentic, relevant, not condescending
```

What to Expect
AI will generate a complete lesson with activities, discussions, and applications appropriate for teenagers. You'll need to adjust based on your group's actual maturity, interests, and cultural context.

Follow-Up Prompts
"Add discussion questions specifically addressing how social media shapes their understanding of identity."
Connecting Biblical truth to teenagers' actual daily experiences makes lessons feel relevant.
"Create a service project component where students practice what we're teaching."
Experiential learning helps abstract concepts become concrete and memorable.
"Suggest how to address students who are skeptical or resistant to this topic."
Good youth ministry creates space for doubt and questions, not just teaching answers.

⌨ **Scripture Check**: Youth lessons must be Biblically faithful while being developmentally appropriate. Don't oversimplify Scripture in ways that create theological misconceptions.

🤔 **Your Discernment Matters**: Teenagers can handle theological depth and honest conversation about hard topics. Don't underestimate their capacity for real faith and deep wrestling.

🔒 **Privacy Reminder**: Create safe spaces for vulnerable sharing, but be clear about mandatory reporting requirements for abuse, self-harm, or danger.

How to Use It
1. Use AI's lesson plan and adapt activities based on your group's size, energy, and interests
2. Ask AI to create a series (4-6 weeks) with connected themes rather than standalone lessons
3. Have AI develop parent communication materials explaining what students are learning

Reflection Question
What question or struggle are your teenagers actually facing right now that this lesson needs to address honestly?

CHAPTER 69
Writing Family Devotional Guides

Our Goal
Create family devotional guides that help parents lead simple, meaningful faith conversations with their children at home.

The Problem
Many parents want to nurture their children's faith but don't know how to start spiritual conversations or lead family devotions. They need resources that are brief, accessible, and actually work with real kids (short attention spans, resistance, varied ages). Creating these guides requires understanding both theology and child development.

Why This Matters
Faith is formed primarily at home, not at church. Family devotionals equip parents as primary spiritual teachers and create rhythms of faith conversation in daily life. When AI helps you create these guides, you can support families without requiring parents to become theologians or skilled teachers.

The Prompt
```
I'm creating a family devotional guide for [season:
weekly, Advent, Lent, summer] designed for families
with children ages [age range: preschool through
elementary, elementary through teens, mixed ages].

Create devotional entries that include:
- Brief Scripture passage (child-accessible version)
- Simple reflection or story connecting Scripture to
family life
- 2-3 discussion questions appropriate for children
- A prayer prompt the whole family can participate in
- Optional activity or practice (blessing, service,
creativity)

Length per devotional: 150-200 words
Tone: Warm, practical, parent-friendly (not preachy)
Family context: [working parents, limited time,
diverse spiritual maturity]
```

What to Expect
AI will generate complete family devotionals with Scripture, reflection, questions, and activities. The content will be developmentally appropriate but generic until you personalize with examples relevant to your congregation's families.

Follow-Up Prompts
"Add specific activities families can do together that reinforce the devotional theme."
Experiential practices help abstract faith concepts become concrete for children.
"Create a version for single-parent families or blended families navigating complex dynamics."
Different family structures need different approaches and sensitivity.
"Adjust this for families with teenagers who might resist 'devotional time.'"
Older kids need different engagement strategies than young children.

Your Discernment Matters: Family devotionals should support parents, not create guilt about one more thing they're failing to do. Keep expectations realistic.

Scripture Check: Ensure Scripture is presented accurately and age-appropriately without distorting meaning or adding details not in the text.

How to Use It
1. Use AI's devotionals and test them with actual families for feedback
2. Ask AI to create seasonal series (Advent, Lent) or topical series (kindness, gratitude, service)
3. Have AI develop parent coaching materials explaining how to lead devotions with resistant or distracted kids

Reflection Question
What realistic rhythm of family devotions would actually be sustainable for the busy families in your congregation?

CHAPTER 70
Creating Vacation Bible School Curriculum

Our Goal
Develop Vacation Bible School curriculum that teaches Biblical stories, creates fun experiences, and serves both churched kids and community children.

The Problem
VBS is one of the biggest outreach events many churches do annually, but creating curriculum from scratch (Bible lessons, crafts, games, music, snacks, decorations) for a week-long program is massive work. Pre-packaged curriculum can be expensive and may not fit your theology or context.

Why This Matters
VBS introduces children to Biblical stories, builds positive church experiences, and connects with families you might not otherwise reach. Well-designed curriculum makes the week run smoothly, volunteers feel equipped, and children encounter God's love. When AI helps you create VBS materials, you can offer quality programming without buying expensive packages or spending months planning.

The Prompt
```
I'm creating Vacation Bible School curriculum for
[age range: preschool through 5th grade] with the
theme of [theme: e.g., God's creation, Jesus'
miracles, heroes of faith, fruit of the Spirit].

Create a 5-day curriculum outline that includes for
each day:
- Daily Bible story and key verse
- Main teaching point (age-appropriate)
- Suggested activities (games, crafts, science
experiments)
- Music or movement activity
- Snack idea connecting to the lesson
- Take-home activity for families

VBS context: [morning/evening program, volunteer-led,
limited budget]
Theological perspective: [Your tradition]
Format: [rotation stations, all-together format,
mixed]
```

What to Expect

AI will generate a complete week-long VBS outline with daily themes, activities, and teaching content. You'll need to adjust based on your space, volunteers, budget, and local children's interests.

Follow-Up Prompts

```
"Add specific craft instructions for Day 3 that
reinforce the lesson about Jesus calming the storm."
```
Detailed activity instructions help volunteers who aren't naturally creative feel equipped.
```
"Suggest outdoor games and activities – we have
limited indoor space." [or vice-versa]
```
Space and resource constraints require adaptation.
```
"Create registration and parent communication
materials explaining the week's schedule and theme."
```
Well-informed parents are more likely to register kids and volunteer.

📖 **Scripture Check**: VBS teaches foundational Biblical stories. Ensure stories are accurate and age-appropriate without adding fictional details or distorting meaning.

🫥 **Your Discernment Matters**: Not all AI-suggested activities are safe, affordable, or practical. Review carefully and test before committing to a week-long plan.

🔒 **Privacy Reminder**: VBS involves children from the community. Have clear policies about photography, collecting emergency / medical information, as well as background checks and training for volunteers.

How to Use It

1. Use AI's curriculum outline and adjust activities based on your budget, space, and volunteer capacity
2. Ask AI to create volunteer training materials and daily lesson scripts
3. Have AI develop follow-up materials for families after VBS ends

Reflection Question

What do you most want visiting children and families to experience about God and church during VBS week?

CHAPTER 71
Developing Mentorship Program Guides and Discussion Prompts

Our Goal
Create mentorship program structures and discussion guides that facilitate meaningful spiritual friendships between mentors and mentees.

The Problem
Mentorship is powerful for discipleship, but most people don't know how to be mentors or what to talk about in mentoring relationships. You want to launch a mentorship ministry, but without structure and resources, mentors feel lost and relationships fizzle. Creating comprehensive mentorship guides requires understanding relational dynamics, spiritual formation, and practical facilitation skills.

Why This Matters
Mentorship personalizes discipleship and creates intergenerational connection. Well-structured programs give mentors confidence, provide conversation frameworks, and help relationships move beyond surface-level friendship into genuine spiritual formation. AI can help you create mentorship resources to launch or refresh your mentoring program.

The Prompt
```
I'm creating a mentorship program guide for [type:
spiritual mentorship, leadership development, new
believer support, youth-to-adult, women's/men's
ministry].

Create mentorship materials that include:
- Program overview and expectations (frequency,
duration, commitments)
- Matching process guidance
- First meeting conversation guide
- 6-8 monthly meeting discussion topics and questions
- Resources for mentors (how to listen, ask
questions, share your story)
- When to seek help or refer to professional
counseling

Mentorship focus: [spiritual formation, leadership
development, life transitions]
Participant context: [ages, spiritual maturity,
church involvement]
```

```
Tone: Practical, encouraging, relationship-focused
```

What to Expect
AI will generate complete mentorship program materials with structure, discussion guides, and mentor training content. You'll need to customize based on your congregation's culture and specific mentorship goals.

Follow-Up Prompts
```
"Add specific discussion questions for Month 3
focused on spiritual practices and disciplines."
```
Thematic progression helps mentoring relationships deepen over time rather than staying surface-level.
```
"Create boundary-setting guidelines for mentors about
appropriate questions, activities, etc."
```
Healthy mentorship requires clear boundaries around time, emotional intimacy, and confidentiality.
```
"Suggest how mentors can identify when mentees need
professional counseling beyond peer mentorship."
```
Mentors need to know their limits and when to refer for professional help.

🧑 **Your Discernment Matters**: Mentorship is relational, not programmatic. Provide structure without making it feel like a class

🔒 **Privacy Reminder**: Establish clear confidentiality expectations and boundaries. Mentees need to know what stays private and what mentors might need to share with leadership.

✝️ **Theological Caution**: Ensure mentorship focuses on spiritual formation, not just advice-giving or fixing people's problems.

How to Use It
1. Use AI's program guide and adjust based on your congregation's relational culture
2. Ask AI to create mentor training materials and sample conversation scripts
3. Have AI develop check-in surveys for mentors and mentees to assess their progress and the value or quality of their time together

Reflection Question
What support or training would help mentors in your congregation feel confident and equipped rather than anxious about "doing it right"?

SECTION 6

ADMINISTRATION & LEADERSHIP

CHAPTER 72
Creating Meeting Agendas

Our Goal
Design clear, efficient meeting agendas that keep church leadership meetings focused, productive, and respectful of people's time.

The Problem
Church meetings often run long, drift off-topic, or leave important decisions unaddressed because of poor agenda planning. You want meetings that accomplish necessary business without exhausting volunteers who already give limited time. Creating agendas that balance information sharing, decision-making, and spiritual formation takes planning you may not have time for.

Why This Matters
Good agendas honor people's time and create space for meaningful work. Well-structured meetings build trust, produce better decisions, and leave people energized rather than drained. AI can help you create clear, comprehensive agendas so you can lead productive meetings without spending hours on preparation.

The Prompt
```
I'm creating an agenda for [meeting type: Church
Council, Finance Committee, Trustees, Staff Meeting]
on [date].

Create an agenda that includes:
- Opening prayer or devotional (5 minutes)
- Approval of previous minutes (5 minutes)
- Reports needing board action [specify which]
- Discussion items with time allocations
- Decision items requiring votes
- Information-only updates
- Closing and adjournment

Meeting length: [60 minutes, 90 minutes, 2 hours]
Key decisions needed: [list major items]
Tone: Efficient, clear, action-oriented
```

What to Expect
AI will generate a structured agenda with time allocations and clear item categorization. You'll need to add specific details about reports, decisions, and your church's actual business.

Follow-Up Prompts
"Add realistic time estimates for each agenda item based on typical discussion length."
Accurate time estimates help meetings stay on schedule and signal which items need more discussion.
"Reorganize the agenda to handle time-sensitive decisions early in the meeting."
Strategic ordering ensures critical decisions get made before people are tired or the meeting runs over.
"Include consent agenda items that can be approved in bulk without discussion."
Consent agendas free up time for items actually requiring conversation and debate.

 Your Discernment Matters: Agendas can't anticipate every conversation that needs to happen. Build in flexibility for emergent issues while maintaining structure.

 Verify This: Ensure all legally required items (financial reports, policy approvals) are included per your denomination's governance requirements.

How to Use It
1. Use AI's agenda template and customize with your church's specific business items
2. Ask AI to create standard agenda templates you can use for recurring meetings
3. Have AI draft background materials or reports referenced in the agenda (ex: "use the attached P&L sheet to create the monthly treasurer's report summarizing income, expenses, budget variances, and comparisons to last year)

Reflection Question
What agenda items could be handled via email or consent agenda to free up meeting time for strategic conversation?

CHAPTER 73
Drafting Church Policies (Safety, Facility Use, Volunteers)

Our Goal
Create clear church policies that protect people, establish expectations, and provide guidance for common situations.

The Problem
Churches need policies for child safety, facility use, volunteer screening, financial controls, conflict resolution, and more. But writing comprehensive policies requires legal awareness, risk management knowledge, and careful language. You know you need policies, but drafting them from scratch feels overwhelming.

Why This Matters
Good policies protect vulnerable people, reduce liability, and provide clarity when conflicts arise. Well-written policies also demonstrate professionalism and care. AI can help you draft policies and create necessary documents without spending weeks on research.

The Prompt
```
I'm creating a church policy for [area: child
protection, facility use, volunteer screening,
financial controls, social media, conflict
resolution].

Draft a policy that includes:
- Purpose and scope
- Clear definitions of terms
- Specific procedures and requirements
- Roles and responsibilities
- Consequences for violations
- Review and update schedule

Church context: [size, state/location for legal
compliance, denominational requirements]
Tone: Clear, professional, protective (not punitive)
Length: 300-400 words
```

What to Expect
AI will generate a well-structured policy with appropriate sections and clear language. You'll need to verify legal compliance, add specific names/roles, and customize for your denomination's requirements.

Follow-Up Prompts
`"Add specific screening requirements complying with [state] law for volunteers working with minors."`
Legal requirements vary by state and must be incorporated accurately.
`"Include reporting procedures when policy violations occur."`
Policies need enforcement mechanisms and clear reporting chains.
`"Create a simplified summary version for volunteers who don't need to read the full policy."`
Accessibility matters – most people need the highlights, not legal language.

⚠ **Verify This**: Church policies must comply with state laws, denominational requirements, and insurance mandates. Have policies reviewed by legal counsel before adoption.

🌐 **Your Discernment Matters**: Policies should protect people, not just protect the institution. Keep vulnerable people at the center of your conversations and policies.

🔒 **Privacy Reminder**: Policies involving background checks or personal information must comply with privacy laws and data protection requirements.

How to Use It
1. Use AI's policy draft as a starting point and have it reviewed by legal counsel
2. Ask AI to create implementation checklists for rolling out new policies or onboarding new staff or volunteers
3. Have AI develop training materials explaining policies to staff and volunteers

Reflection Question
What incident or vulnerability does this policy need to prevent, and does the language actually accomplish that goal?

CHAPTER 74
Writing Job Descriptions for Staff and Volunteers

Our Goal
Create clear job descriptions that define roles, set expectations, and help people understand what success looks like.

The Problem
Vague job descriptions lead to confusion, disappointment, and conflict. Staff and volunteers need to know what they're responsible for, who they report to, and how they'll be evaluated. Writing comprehensive job descriptions that balance detail with flexibility takes HR expertise most pastors and lay people don't have.

Why This Matters
Clear job descriptions prevent misunderstandings, enable accountability, and help people thrive in their roles. They also protect both employee and employer when performance issues arise. When AI helps you write job descriptions, you can create professional HR documents without hiring consultants.

The Prompt
```
I'm writing a job description for [position: youth
pastor, administrative assistant, worship leader,
nursery coordinator, facilities manager, volunteer
coordinator].

Create a job description that includes:
- Position title and reporting structure
- Purpose/summary of the role
- Key responsibilities (5-8 primary duties)
- Required qualifications and skills
- Desired qualifications
- Time commitment or hours
- Compensation range (if applicable)

Church context: [size, ministry focus, staff
structure]
Position type: [full-time, part-time, volunteer]
Tone: Professional, clear, invitational
```

What to Expect
AI will generate a professional job description with standard HR structure and clear language. You'll need to customize with your church's specific expectations, compensation, and culture.

Follow-Up Prompts
```
"Add specific metrics or outcomes that define success
in this role."
```
Measurable expectations help people know when they're succeeding and provide basis for evaluation.
```
"Adjust the qualifications to prioritize character
and calling over professional credentials."
```
Ministry roles often value spiritual maturity and relational skills differently than corporate jobs.
```
"Create a volunteer version of this job description
with appropriate time commitment expectations."
```
Volunteers need clarity but different framing than paid staff positions.

⚠ **Verify This**: Job descriptions must comply with employment law, wage/hour regulations, and equal opportunity requirements. Consult legal or HR professionals.

🧭 **Your Discernment Matters**: Job descriptions shouldn't be so rigid they prevent flexibility, or so vague they provide no accountability.

How to Use It
1. Use AI's job description as a foundation and customize with your church's specific culture and expectations
2. Ask AI to create interview questions aligned with the job description
3. Have AI develop evaluation rubrics based on the job description's key responsibilities

Reflection Question
If someone succeeded wildly in this role, what would be different in your church's ministry a year from now?

CHAPTER 75
Creating Volunteer Recruitment and Onboarding Materials

Our Goal
Develop recruitment materials and onboarding processes that help volunteers find meaningful roles and get equipped to serve well.

The Problem
You need volunteers, but generic "we need help" announcements don't work. People want to know what they're signing up for, what training they'll receive, and whether their service will matter. Creating compelling recruitment materials and thorough onboarding processes for multiple ministry areas is time-consuming.

Why This Matters
Good recruitment connects people's passions with ministry needs. Thorough onboarding helps volunteers feel equipped, valued, and likely to continue serving. When AI helps you create these materials, you can recruit and equip volunteers systematically rather than desperately.

The Prompt
```
I'm creating volunteer recruitment and onboarding
materials for [ministry area: children's ministry,
hospitality team, worship tech, food pantry, care
ministry].

Create materials that include:
- Compelling recruitment description (100 words)
explaining the role and impact
- Time commitment and expectations clearly stated
- Required qualifications or screening
- Onboarding checklist (training, background checks,
orientation)
- First-day welcome and setup
- Ongoing support and appreciation plan

Ministry context: [size, volunteer culture, existing
team]
Volunteer type: [weekly commitment, occasional,
seasonal, one-time]
Tone: Invitational, clear about expectations,
grateful
```

What to Expect
AI will generate recruitment copy, onboarding checklists, and support plans. You'll need to customize with your church's specific requirements, training resources, and volunteer coordinator contact information.

Follow-Up Prompts
"Add specific stories of volunteers who found meaning and community through this ministry."
Real testimonies make recruitment compelling and help people envision themselves serving.
"Create a timeline showing when volunteers need to complete each onboarding step."
Clear timelines help volunteers and coordinators stay organized and accountable.
"Develop a 30-,60-,90-day check-in plan for new volunteers."
Regular check-ins during the first months prevent volunteers from feeling lost or quitting quietly.

🔒 **Privacy Reminder**: Volunteer screening must comply with privacy laws, especially background checks for those working with vulnerable populations.

⚠ **Verify This**: Ensure onboarding includes all legally required training (child protection, harassment prevention) and church-required clearances.

🙂 **Your Discernment Matters**: Don't oversell volunteer roles or hide challenges. Honesty about both rewards and difficulties leads to better volunteer retention.

How to Use It
1. Use AI's recruitment materials and personalize with actual volunteer stories from your church
2. Ask AI to create role-specific onboarding checklists for different ministry areas
3. Have AI develop volunteer appreciation materials and milestone recognition

Reflection Question
What would make someone feel genuinely valued and equipped in this volunteer role rather than just filling a slot?

CHAPTER 76
Developing Volunteer Schedules for Recurring Ministries

Our Goal
Create sustainable volunteer schedules for recurring ministries that prevent burnout, ensure coverage, and honor people's availability.

The Problem
Scheduling volunteers for weekly or monthly ministries (ushers, greeters, nursery, coffee hour, tech team) is administrative burden. You're constantly filling gaps, managing conflicts, and trying to remember who's available when. Without a good system, scheduling becomes reactive crisis management.

Why This Matters
Predictable schedules help volunteers plan their lives and serve sustainably. Well-designed rotation systems prevent burnout and ensure consistent ministry quality. When AI helps you create scheduling systems, you can move from constant scrambling to proactive management.

The Prompt
```
I'm creating a volunteer schedule for [ministry:
Sunday greeters, nursery workers, worship tech team,
coffee hour hosts, ushers] with [number] volunteers
rotating on [frequency: weekly, monthly, quarterly].

Design a scheduling system that includes:
- Rotation pattern ensuring fair distribution
- Coverage for all needed dates
- Backup/substitute process when conflicts arise
- Advance notice period (how far ahead people know
their dates)
- Communication method (email, text, app)
- How to request time off or find substitutes

Ministry context: [service frequency, team size,
special event considerations]
Volunteer preferences: [some prefer regular rotation,
others occasional]
```

What to Expect
AI will generate a rotation schedule with fair distribution and clear coverage. You'll need to input actual volunteer names, availability constraints, and blackout dates.

Follow-Up Prompts
```
"Add holiday coverage plan accounting for volunteers
traveling or hosting family members from out of
town."
```
Holiday scheduling requires extra planning since many volunteers have conflicts.
```
"Create a communication template for reminding
volunteers of their upcoming schedule."
```
Automated reminders reduce no-shows and last-minute scrambling.
```
"Suggest how to build in flexibility for volunteers
with unpredictable work schedules."
```
Rigid schedules don't work for everyone – some need more flexibility to serve sustainably.

 Your Discernment Matters: Scheduling systems should serve people, not become burdensome administrative machinery. Keep it simple enough to maintain.

 Verify This: For ministries requiring background checks or training, ensure scheduling only includes cleared volunteers.

How to Use It
1. Use AI's rotation pattern and populate with your actual volunteer names and availability
2. Ask AI to create different scheduling models (weekly rotation, monthly teams, on-call lists) and pick what works best
3. Have AI develop volunteer communication templates for schedule reminders and substitute requests

Reflection Question
What scheduling approach would actually reduce your administrative burden while honoring volunteers' need for predictability and flexibility?

CHAPTER 77
Planning Multi-Day Events (Retreats, Conferences, VBS)

Our Goal
Create comprehensive plans for multi-day church events that coordinate logistics, programming, volunteers, and budgets without losing your mind.

The Problem
Multi-day events like retreats, conferences, or VBS involve coordinating facilities, meals, programming, childcare, transportation, and dozens of volunteers. The logistics are overwhelming, and missing one detail can derail the whole event. Planning without a systematic approach leads to stress and forgotten essentials.

Why This Matters
Well-planned events create transformative experiences. Poor planning creates chaos that undermines ministry impact and exhausts volunteers. When AI helps you plan multi-day events, you can think systematically through complex logistics without forgetting critical details.

The Prompt
```
I'm planning a [event type: church retreat,
leadership conference, VBS, mission trip] for
[number] participants from [dates].

Create a planning framework that includes:
- Timeline working backward from event date (3 months
out, 2 months, 1 month, 2 weeks, 1 week)
- Budget categories and estimated costs
- Logistics checklist (facility, food,
transportation, childcare, supplies)
- Programming schedule (sessions, free time, meals,
worship)
- Volunteer roles and responsibilities
- Communication plan for participants
- Risk management and emergency plans

Event context: [location, participant demographics,
budget constraints]
Format: [residential retreat, day program, off-site,
on-campus]
```

What to Expect

AI will generate a comprehensive planning framework with timelines, checklists, and budget categories. You'll need to customize with your specific event details, venue requirements, and actual costs.

Follow-Up Prompts

```
"Add specific volunteer job descriptions for each
role and how many people we need."
```
Clear role definitions help recruitment and ensure adequate coverage.
```
"Create a packing list for participants so they know
what to bring."
```
Detailed participant communication reduces confusion and last-minute questions.
```
"Develop contingency plans for bad weather, illness,
or emergency situations."
```
Multi-day events need backup plans for when things go wrong.

⚠ **Verify This**: Check facility contracts, insurance requirements, food safety regulations, and transportation licensing before finalizing plans.

🔒 **Privacy Reminder**: Collect necessary medical information and emergency contacts for all participants, especially minors. Store securely and share only with authorized staff.

🙂 **Your Discernment Matters**: Don't over-program. Multi-day events need downtime, free conversation, and space for God's Spirit to work outside scheduled activities.

How to Use It

1. Use AI's planning framework 3-6 months before your event and customize the planning / training / rollout timeline based on your preparation capacity
2. Ask AI to create different budget scenarios (minimal, moderate, ideal) to guide decision-making
3. Use AI to help you write grants to get funding for your event
4. Have AI develop post-event evaluation surveys and volunteer debriefs

Reflection Question

What's the one thing that would most enhance this event's spiritual impact, and are you allocating enough time, space, and resources to make it happen?

CHAPTER 78
Creating Event Checklists and Timelines

Our Goal
Develop detailed event checklists and timelines that ensure nothing gets forgotten and everything happens on schedule.

The Problem
Events fail because of forgotten details – no one ordered coffee, the sound system isn't set up, volunteers don't know their roles, or the timeline falls apart. You can't hold every detail in your head, but creating comprehensive checklists for every event takes significant planning time.

Why This Matters
Checklists free your brain from remembering details so you can focus on people and ministry. Good timelines keep events flowing smoothly and signal when things are running behind. When AI helps you create these tools, you can lead events confidently without constant anxiety about what you might have forgotten.

The Prompt
```
I'm creating a checklist and timeline for [event
type: Sunday worship, special service, community
dinner, fundraising event, memorial service].

Create event materials that include:
- Master checklist organized by category (setup,
technology, hospitality, program, cleanup)
- Timeline showing when each task needs to be
completed (weeks before, days before, day-of hour-by-
hour)
- Responsible parties for each task
- Setup and teardown instructions
- Volunteer assignments and briefing points

Event context: [size, complexity, frequency]
Venue: [sanctuary, fellowship hall, outdoor, off-
site]
Format: Checklist and hour-by-hour timeline
```

What to Expect

AI will generate comprehensive checklists and realistic timelines. You'll need to customize with your venue's specific setup requirements and add names of responsible volunteers.

Follow-Up Prompts

"Add specific technology setup steps (sound check, slides loaded, livestream tested)."

Technical details are often where events break down – explicit checklists prevent failures.

"Create a condensed day-of timeline for volunteers who just need to know their specific tasks."

Volunteers don't need the master timeline – they need clear instructions for their role.

"Include contingency items for common problems (backup supplies, weather alternatives, tech failures, leader / performer gets sick)."

Planning for likely failures prevents panic when things go wrong.

Your Discernment Matters: Checklists help but can't anticipate everything. Build in buffer time for the unexpected.

Verify This: Walk through your venue with the checklist to ensure it matches your actual space and equipment.

How to Use It

1. Use AI's checklist and timeline as templates, then customize for your specific venue and event
2. Ask AI to create role-specific checklists for different volunteer positions
3. Have AI develop post-event debriefs identifying what worked and what needs adjustment for next time

Reflection Question

What task or detail have you forgotten at past events that needs to be on this checklist?

CHAPTER 79
Writing Budget Narratives and Stewardship Explanations

Our Goal
Create clear budget narratives that explain how church funds are used and why stewardship matters, making financial information accessible and inspiring.

The Problem
Church budgets are full of line items and numbers that mean nothing to most people. You need to communicate financial information in ways that build trust, inspire giving, and show ministry impact. Writing compelling stewardship narratives that connect dollars to discipleship takes skill and time.

Why This Matters
Transparent, story-driven financial communication builds trust and generous giving. When people understand how their contributions fund actual ministry, they give more freely and faithfully. When AI helps you write budget narratives, you can make financial information compelling without becoming a professional grant writer.

The Prompt
```
I'm writing a budget narrative explaining our
church's [annual budget, capital campaign, special
fundraising appeal] totaling [amount].

[If possible, attach a PDF or Excel file of your
current budget so AI can use your actual numbers.]

Create narrative content (300-400 words) that:
- Opens with vision and mission (why we do what we
do)
- Explains major budget categories in ministry impact
terms, not accounting jargon
- Shares specific stories of lives changed through
funded ministries
- Shows percentage breakdowns visually (worship,
missions, staff, facilities, etc.)
- Closes with invitation to generous partnership

Church context: [size, ministries, giving culture]
Budget focus: [operating budget, building project,
mission support]
```

```
Audience: Congregation members and potential donors
Tone: Transparent, grateful, inspiring (not guilt-
driven)
```

What to Expect
AI will generate budget narrative connecting finances to ministry impact with stories and vision. You'll need to add your church's actual numbers, percentages, and real ministry stories.

Follow-Up Prompts
```
"Add specific examples showing what different giving
levels support ($50/month, $200/month, etc.)."
```
Concrete giving examples help people envision their personal contribution's impact.
```
"Create a visual breakdown showing percentage
allocation across major budget categories."
```
Visual representations make complex budgets accessible to non-accountants.
```
"Include frequently asked questions about budget
priorities and financial decisions."
```
Proactive FAQ addresses common concerns and builds transparency.

⚠ **Verify This**: All financial figures, percentages, and ministry impact claims must be accurate and verifiable. Transparency builds trust.

😊 **Your Discernment Matters**: Budget narratives should inspire generosity, not manipulate through guilt or pressure. Focus on joyful partnership in God's work and specific outcomes / impacts.

How to Use It
1. Use AI's narrative framework and insert your church's actual budget numbers and ministry stories
2. Ask AI to create different versions for different audiences (members, donors, denominational reports)
3. Have AI develop visual infographics or charts accompanying the narrative

Reflection Question
If someone asked "Where does my money actually go?" could they find a clear, compelling answer in this budget narrative?

CHAPTER 80
Drafting Grant Proposals for Ministry Funding

Our Goal
Write compelling grant proposals that secure funding for ministry projects from foundations, denominational grants, or community funders.

The Problem
Grant funding could expand your ministry, but writing competitive proposals requires understanding funder priorities, demonstrating impact, creating budgets, and writing persuasively. Most pastors have never written grants and don't know where to start.

Why This Matters
Grant funding enables ministries you couldn't afford otherwise. Well-written proposals open doors to resources for community programs, building projects, or special initiatives. When AI helps you write grant proposals, you can pursue funding opportunities without hiring grant writers or spending weeks on applications.

The Prompt
```
I'm writing a grant proposal to [funder name or type]
requesting [amount] for [project: community program,
building renovation, new ministry initiative,
equipment].

[If possible, attach a PDF of the grant application
form so AI can better understand what is needed.]

Create a grant proposal that includes:
- Executive summary (project overview and funding
request)
- Need statement (problem being addressed with
data/evidence)
- Project description (activities, timeline,
participants served)
- Goals and measurable outcomes
- Budget and budget justification
- Organizational capacity (why we're qualified to do
this)
- Sustainability plan (how project continues after
grant ends)
```

```
Project context: [community need, people served,
timeline]
Funder priorities: [if known: youth, poverty,
education, health, etc.]
Length: 400-500 words (adjust sections as needed)
```

What to Expect
AI will generate a complete grant proposal with all standard sections and persuasive language. You'll need to add specific data about your community, actual budget numbers, and evidence of need.

Follow-Up Prompts
```
"Add specific demographic data about the population
we're serving and the community need."
```
Funders want evidence-based proposals showing genuine need, not just good ideas.
```
"Strengthen the outcomes section with specific,
measurable metrics for success."
```
Vague goals like "help people" don't win grants – measurable outcomes do.
```
"Include letters of support or partnership from other
organizations we're collaborating with."
```
Collaborative projects often score higher than solo church initiatives.

⚠ **Verify This**: All data, statistics, budget figures, and partnership claims must be accurate and verifiable. Grant fraud has serious legal consequences.

🧭 **Your Discernment Matters**: Only pursue grants aligned with your actual mission. Don't let funding opportunities distort your ministry priorities.

How to Use It
1. Use AI's proposal as your foundation and strengthen with local data and real community stories
2. Ask AI to create different proposal versions customized for different funders' priorities
3. Have AI develop evaluation plans and reporting templates funders will require

Reflection Question
Does this project genuinely serve community need, or are you shaping ministry around available funding?

CHAPTER 81
Reviewing Contracts and Lease Agreements

Our Goal
Get a second opinion on contracts or lease agreements, identifying potential concerns, unclear terms, or unfavorable clauses before you sign.

The Problem
Pastors aren't lawyers. When your church is asked to sign a facility lease, vendor contract, or partnership agreement, you're responsible for protecting the congregation's interests – but legal language is confusing and intimidating. You don't want to miss a problematic clause, but hiring an attorney for every small contract isn't realistic.

Why This Matters
Bad contracts can lock your church into unfavorable terms, create liability issues, or cost thousands of dollars to exit. A careful contract review protects your congregation from legal and financial harm. AI can't provide legal advice, but it can help you spot red flags that warrant a lawyer's attention or questions you should ask before signing.

The Prompt
```
I'm a pastor reviewing [type of contract: facility
rental agreement, landscaping service contract,
equipment lease] for our church.

Here is the contract text:
[Paste contract text or key sections]
[If possible, attach a PDF of the entire contract,
your annual budget and any other related documents.]

Please review this contract and:
1. Identify any clauses that seem unusual, one-sided,
or potentially problematic
2. Flag unclear terms or definitions that need
clarification
3. Note any financial obligations or penalties
4. Highlight liability or indemnification clauses
5. Suggest questions I should ask the other party
before signing

Context: Our church is [size, budget, context]
Provide your analysis in clear, non-legal language.
```

What to Expect

AI will identify concerning clauses, flag confusing language, and suggest clarifying questions. It will explain legal terms in plain English and highlight financial or liability risks. It cannot, however, provide definitive legal advice or tell you whether to sign.

Follow-Up Prompts

"Are there standard clauses missing from this contract that should be included?"
This checks for omissions, not just problems – sometimes what's NOT in a contract matters as much as what is.
"Explain the indemnification clause in simpler terms – what does it actually mean for our church?"
AI can translate legal jargon into everyday language so you understand what you're agreeing to.
"Draft 3-10 questions I should ask the vendor based on your concerns."
This moves you from analysis to action – knowing what to ask before you sign.

⚠ **Verify This**: AI is not a lawyer and cannot provide legal advice. Use AI's review as a starting point to identify concerns, but always consult an attorney for contracts involving significant financial commitments, long-term obligations, or liability risks. AI can help you know *when* to call a lawyer and make those conversations more focused and productive.

🤔 **Your Discernment Matters**: Even with AI review, trust your instincts. If something feels wrong or too risky, don't sign without professional legal review.

How to Use It

1. Use AI's analysis to understand the contract's key terms and risks
2. Ask AI to identify and prioritize which concerns are serious enough to warrant legal counsel
3. Have AI draft clarifying questions to ask the other party before signing

Reflection Question

Based on AI's review, is this contract something you feel comfortable signing after clarification, or does it raise enough concerns that you should involve legal counsel or reconsider the agreement entirely?

CHAPTER 82
Creating Strategic Planning Documents and Vision Statements

Our Goal
Develop strategic planning documents and vision statements that articulate where your church is going and how you'll get there.

The Problem
Churches drift without clear vision and strategic direction. You know you need a plan beyond "keep doing what we're doing," but facilitating strategic planning processes and writing vision documents requires leadership skills and strategic thinking that feels overwhelming alongside pastoral responsibilities.

Why This Matters
Strategic plans align leadership, focus resources, and create accountability for moving toward shared goals. Vision statements inspire and unite congregations around common purpose. When AI helps you create these documents, you can lead strategic planning without expensive consultants or endless committee meetings.

The Prompt
```
I'm creating strategic planning documents for our
church. [Insert name, denomination, city, state.]

Create materials that include:
- Vision statement (1 sentence describing why we
exist) based on Matthew 22:37-40
- Mission statement (how we live into our vision)
based on Matthew 28:18-20
- Core values (3-5 values that shape our decisions)
based on 1 Corinthians 13:13
- Strategic goals for [3-year/5-year] timeframe
- Key initiatives under each goal
- Success metrics and accountability

Church context: [size, strengths, challenges,
community demographics and economic context]
Current reality: [what's working, what needs change]
Leadership input: [themes from leadership
conversations]
Tone: Inspiring, clear, action-oriented
```

What to Expect
AI will generate vision / mission statements, strategic goals, and planning frameworks. You'll need to ground these in your church's actual DNA, leadership input, and community calling.

Follow-Up Prompts
```
"Make the vision statement simple enough for a smart
6-year-old to understand and share with their
friends."
```
Vision statements should be simple, clear, and distinctively yours.
```
"Add specific measurable outcomes for each strategic
goal so we know if we're succeeding."
```
Strategy without accountability becomes wishful thinking.
```
"Create an implementation timeline showing when we'll
work on each initiative."
```
Good plans include realistic sequencing – you can't do everything at once.

⚙ **Your Discernment Matters**: Strategic plans must emerge from genuine leadership discernment and congregational input, not just AI-generated documents. Use AI to organize thinking, not replace it.

✝ **Theological Caution**: Church vision must be rooted in Biblical mission and calling, not just institutional growth or organizational success.

How to Use It
1. Use AI's framework after gathering leadership input and congregational feedback
2. Ask AI to create presentation materials for sharing the plan with the congregation
3. Have AI develop annual review templates tracking progress toward strategic goals

Reflection Question
If you accomplish everything in this strategic plan, what will be fundamentally different about your church's ministry and impact in your community?

CHAPTER 83
Writing Annual Reports for Congregation or Conference

Our Goal
Create comprehensive annual reports that celebrate ministry impact, provide transparency, and inspire continued partnership.

The Problem
Annual reports require compiling data from multiple ministries, writing compelling narratives, and presenting financial information clearly. You need reports that inform without boring people, celebrate without bragging, and inspire without manipulating. Creating these documents takes significant time and writing skill.

Why This Matters
Annual reports demonstrate accountability, celebrate God's faithfulness, and cast a vision for your church's future. They educate new members, satisfy denominational requirements, and provide a historical record. When AI helps you write annual reports, you can produce professional documents without spending weeks on compilation and writing.

The Prompt
```
I'm writing an annual report for [audience:
congregation, denominational conference, donors]
covering [year].

[If possible, attach PDFs of your annual budget,
committee reports, and any other related documents.]

Create report content that includes:
- Opening message from the pastor highlighting the
year's themes
- Ministry area summaries (worship, education,
mission, care)
- Key statistics and metrics (attendance, giving,
participation)
- Stories of transformation and impact
- Financial summary with transparency
- Looking ahead: vision for next year
- Gratitude and closing

Church context: [size, major initiatives this year,
challenges faced]
Report format: [printed booklet, digital PDF, slides]
```

```
Tone: Grateful, transparent, inspiring
Length: 400-500 words of narrative content
```

What to Expect

AI will generate report narratives with appropriate structure and tone. You'll need to add actual statistics, real ministry stories, financial data, and specific names/details from your year.

Follow-Up Prompts

```
"Add specific stories from our food pantry ministry
showing lives changed through this work."
```
Stories make statistics meaningful and help people see ministry impact beyond numbers.
```
"Include challenges we faced this year and how we
responded, not just successes."
```
Honest reporting builds trust and shows resilience through difficulty.
```
"Create a visual timeline or infographic highlighting
major events throughout the year."
```
Visual elements make reports engaging and help readers quickly grasp the year's arc.

⚠ **Verify This**: All statistics, financial figures, and ministry claims must be accurate and verifiable. Annual reports are official records.

🤔 **Your Discernment Matters**: Balance celebration with honesty. Reports that only highlight success without acknowledging challenges feel manipulative.

How to Use It

1. Use AI's report framework and populate with your church's actual data and stories
2. Ask AI to create different versions for different audiences (internal congregation vs. denominational report)
3. Have AI develop visual layouts and design suggestions for printed or digital formats

Reflection Question

What's the one ministry story from this year that best captures God's work in your congregation?

CHAPTER 84
Developing Leadership Training Materials

Our Goal
Create training materials that equip church leaders with skills, knowledge, and spiritual formation for their roles.

The Problem
Church leaders need training in areas like conflict resolution, budget management, spiritual discernment, and team leadership. You want to develop leaders without sending everyone to expensive conferences or spending months creating curriculum. Quality leadership development requires instructional skills you may not have.

Why This Matters
Well-trained leaders make better decisions, handle conflict constructively, and serve longer with less burnout. Leadership development is discipleship at its best. When AI helps you create training materials, you can offer substantive leadership formation without extensive curriculum development.

The Prompt
```
I'm creating leadership training materials for
[audience: church board members, committee chairs,
ministry leaders, staff] on [topic: spiritual
discernment, conflict management, financial
stewardship, team building, meeting facilitation].

Create training content that includes:
- Learning objectives (what leaders will know/be able
to do)
- Biblical/theological foundation for this leadership
skill
- Practical skills or frameworks
- Case studies or scenarios for practice
- Discussion questions for group processing
- Resources for continued learning

Leadership context: [volunteer vs. staff, experience
level, time available]
Format: [workshop, retreat, video series, reading
group]
Length: [1-hour session, half-day, full-day retreat]
```

What to Expect
AI will generate complete training content with objectives, teaching material, activities, and resources. You'll need to customize with scenarios relevant to your church's actual leadership challenges.

Follow-Up Prompts
"Add specific case studies based on common conflicts or challenges our leaders face."
Real-world scenarios make training practical and help leaders apply concepts immediately.
"Include Scripture-based spiritual formation components that link to the skills training."
Christian leadership development must include prayer, discernment, and character formation.
"Create a follow-up plan for ongoing learning and accountability after initial training."
One-time training rarely sticks – leaders need continued support and skill reinforcement.

 Your Discernment Matters: Leadership training must be grounded in your church's actual culture and challenges, not generic corporate leadership principles.

 Theological Caution: Christian leadership is servant leadership. Ensure training reflects Biblical models, not just business best practices.

How to Use It
1. Use AI's training content and adapt case studies to your church's real leadership scenarios
2. Ask AI to create different training modules for different leadership roles
3. Have AI develop assessment tools measuring leadership growth over time

Reflection Question
What leadership skill or capacity, if developed in your current leaders, would most strengthen your church's ministry?

CHAPTER 85
Creating Organizational Charts and Ministry Structures

Our Goal
Design clear organizational charts that show reporting relationships, ministry structures, and how different teams connect.

The Problem
As churches grow, ministry structure becomes complex and confusing. People don't know who's responsible for what, who reports to whom, or how decisions get made. You need organizational clarity, but creating charts that accurately reflect your structure without becoming bureaucratic is challenging.

Why This Matters
Clear organizational structures prevent conflicts, enable accountability, and help people know where they fit. Good org charts also identify gaps, redundancies, or unclear reporting relationships. When AI helps you create these documents, you can bring clarity to complex structures without hiring organizational consultants.

The Prompt
```
I'm creating an organizational chart for our church
showing [scope: full church structure, staff only,
volunteer leadership, specific ministry area].

Design a structure that shows:
- Reporting relationships (who reports to whom)
- Key roles and positions
- Ministry areas and their leaders
- Decision-making flow
- Support/dotted-line relationships where relevant

Church context: [size, staff structure, governance
model]
Format: [hierarchical chart, circular/collaborative
model, matrix]
Complexity: [simple overview vs. detailed
comprehensive]
```

What to Expect

AI will generate organizational structure suggestions with clear reporting lines. You'll need to customize with actual names, titles, and your church's specific governance model.

Follow-Up Prompts

"Adjust this to reflect our collaborative team model rather than top-down hierarchy."

Different leadership philosophies require different organizational representations.

"Show how volunteer ministry leaders connect to staff oversight and support."

Volunteer structures often have different reporting relationships than staff.

"Identify any gaps or redundancies in this structure that need addressing."

Good org charts reveal structural problems that need fixing.

Your Discernment Matters: Org charts can feel institutional and impersonal. Frame them as tools for clarity, not bureaucratic control.

Theological Caution: Church structure should reflect Biblical models of servant leadership and shared ministry, not just corporate hierarchy.

How to Use It

1. Use AI's structural suggestions and customize with your actual roles and relationships
2. Ask AI to create narrative descriptions explaining the chart for people who prefer written explanations
3. Have AI identify structural improvements or realignments needed

Reflection Question

Does this organizational structure actually reflect how decisions get made and ministry happens, or just how you wish it worked?

CHAPTER 86
Writing Welcome Packets for New Staff or Volunteers

Our Goal
Create comprehensive welcome packets that help new staff or volunteers get oriented, feel valued, and succeed in their roles.

The Problem
New staff and volunteers need to know church culture, policies, key contacts, and practical details (parking, building access, schedules). Orienting people verbally is inconsistent and inefficient. You need written materials that provide thorough orientation without overwhelming newcomers.

Why This Matters
Good onboarding helps people succeed and feel welcomed. Welcome packets reduce anxiety, answer common questions, and demonstrate professionalism. When AI helps you create these materials, you can provide consistent, quality orientation for everyone who joins your team.

The Prompt
```
I'm creating a welcome packet for [audience: new
staff members, ministry volunteers, committee
members].

Create packet content that includes:
- Welcome letter from pastor
- Vision, mission, and values
- Organizational overview and key contacts
- Practical information (building access, parking,
supplies, technology)
- Policies and procedures relevant to this role
- Frequently asked questions
- Next steps and orientation schedule

Church context: [size, culture, denominational
background]
Role type: [paid staff, volunteer, leadership]
Format: [printed folder, digital PDF, online portal]
```

What to Expect
AI will generate comprehensive welcome packet content with all standard sections. You'll need to add specific names, phone numbers, building details, and policies relevant to your church.

Follow-Up Prompts
```
"Add a staff/volunteer directory with photos and
contact information."
```
Putting faces to names helps new people connect and know who to ask for help.
```
"Include a glossary of church-specific terms,
acronyms, or insider language."
```
Every church has jargon that insiders forget is confusing to newcomers.
```
"Create a 30-day orientation checklist showing what
new people should do their first month."
```
Structured onboarding prevents people from feeling lost or overlooked.

⚠ **Verify This**: Keep all contact information, policies, and procedures current. Outdated welcome packets create confusion.

🤔 **Your Discernment Matters**: Welcome packets should feel hospitable, not bureaucratic. Balance necessary information with warmth.

How to Use It
1. Use AI's welcome packet content and customize with your actual church details and culture
2. Ask AI to create different versions for different roles (staff, volunteers, committee leaders)
3. Have AI develop orientation schedules and first-day agendas accompanying the packet

Reflection Question
What information or welcome gesture would have most helped you feel oriented and valued when you started at this church?

CHAPTER 87
Drafting Denominational Reports and Compliance Documents

Our Goal
Complete required denominational reports and compliance documents accurately and on time without spending days on paperwork.

The Problem
Denominations require annual reports with statistics, narratives, financial data, and programmatic information. These reports are tedious, time-consuming, and often confusing. You want to fulfill denominational obligations without bureaucratic burdens overwhelming your actual ministry.

Why This Matters
Denominational reports maintain accountability, inform resource allocation, and connect local churches to broader mission. Timely, accurate reporting demonstrates good stewardship and keeps your church in good standing. When AI helps you draft these documents, you can fulfill requirements efficiently.

The Prompt
```
I'm completing a denominational report for
[denomination] requiring [type of information:
membership statistics, financial data, program
participation, narrative descriptions].

[If possible, attach PDFs of your denominational
forms as well as PDFs of your annual budget, P&L
sheet, committee reports, etc. as needed.]

Create report content that includes:
- Required statistical data (I'll provide actual
numbers)
- Narrative descriptions of ministry programs
- Financial summary in requested format
- Significant events or changes this year
- Responses to specific denominational questions
- Future plans or anticipated needs

Denomination: [UMC, Presbyterian, Baptist, etc.]
Report purpose: [annual conference, accountability,
resource planning]
Format: [online form, written document, presentation]
```

What to Expect

AI will generate report narratives and help organize required information clearly. You'll need to input actual statistics, financial data, and denominational-specific requirements.

Follow-Up Prompts

```
"Format this data according to the specific
categories our denomination requires."
```
Denominational reports often have very specific formatting and category requirements.
```
"Add narrative explanation for why our
attendance/giving numbers changed significantly this
year."
```
Context helps denominational leaders understand the numbers, not just see them.
```
"Create a checklist ensuring I've completed all
required sections before submission."
```
Missing required fields causes reports to be rejected or marked incomplete.

⚠ **Verify This**: All statistics, financial figures, and membership data must be accurate. Denominational reports are official records with accountability implications.

🫤 **Your Discernment Matters**: Don't let reporting requirements distort your actual ministry priorities. Report honestly, even when numbers aren't impressive.

How to Use It

1. Use AI to draft narrative sections or an outline while you gather required statistical data
2. Ask AI to create annual reporting calendars reminding you of deadlines throughout the year
3. Have AI develop internal data collection systems making annual reporting easier

Reflection Question

What does this required reporting reveal about your ministry's actual priorities and impact versus what you thought you were emphasizing?

CHAPTER 88
Creating Succession Planning Documents

Our Goal
Develop succession plans that ensure continuity when pastors, staff, or key leaders transition.

The Problem
Pastoral transitions, staff departures, or volunteer leader changes can create chaos without planning. You need documented processes, knowledge transfer systems, and interim plans. Creating succession documentation may feel morbid or presumptuous, but failing to plan creates crisis when transitions inevitably happen.

Why This Matters
Good succession planning protects congregations during leadership transitions. Documented processes preserve institutional knowledge, reduce anxiety, and ensure ministry continuity. When AI helps you create these documents, you can plan responsibly rather than trying to cram the handoff into your last week in the office.

The Prompt
```
I'm creating a succession plan for [role: senior
pastor, staff position, key volunteer leadership,
myself].

Create succession planning documents [or a checklist
with bullet points] that include:
- Critical knowledge and relationships this role
holds
- Key responsibilities and how they'd be covered
during transition
- Important contacts, passwords, and institutional
knowledge
- Timeline for transition process
- Interim leadership structure
- How to communicate transition to
congregation/stakeholders

Context: [church size, governance structure,
denominational process]
Transition scenario: [planned retirement, unexpected
departure, emergency]
Tone: Professional, forward-thinking, not anxious
```

What to Expect
AI will generate comprehensive succession planning frameworks with knowledge transfer checklists and transition timelines. You'll need to add specific details, contacts, and your church's actual governance requirements.

Follow-Up Prompts
```
"Add specific institutional knowledge I hold that
isn't documented anywhere else."
```
Succession planning surfaces critical knowledge that exists only in someone's head.
```
"Create communication templates for announcing
transition to different stakeholder groups."
```
How you announce transitions matters – different audiences need different information.
```
"Include emergency succession plan of essential
functions if I become suddenly unable to serve."
```
Emergency planning ensures continuity even in worst-case scenarios.

😌 **Your Discernment Matters**: Succession planning isn't about being replaceable – it's about being responsible. Good leaders plan for continuity.

🔒 **Privacy Reminder**: Succession documents contain sensitive information (passwords, contacts, institutional knowledge). Store securely and limit access appropriately.

How to Use It
1. Use AI's succession framework and document your actual critical knowledge and relationships
2. Ask AI to create different succession scenarios (planned vs. emergency) with appropriate timelines
3. Have AI develop knowledge transfer checklists ensuring nothing critical gets lost

Reflection Question
If you left tomorrow, what critical institutional knowledge or key relationships would leave with you, and how can you document or handover that before the transition happens?

CHAPTER 89
Writing Performance Review Templates for Staff

Our Goal
Create fair, constructive performance review templates that help staff grow, celebrate strengths, and address areas needing development.

The Problem
Performance reviews are uncomfortable for most pastors. You want to encourage staff, provide honest feedback, and set clear goals, but formal evaluation processes feel corporate or judgmental. You need review systems that honor both accountability and grace.

Why This Matters
Good performance reviews help staff succeed, provide documentation for employment decisions, and create accountability for growth. Fair evaluation systems protect both employee and employer. When AI helps you create review templates, you can conduct professional evaluations without HR expertise.

The Prompt
```
I'm creating a performance review template for
[position: associate pastor, administrative staff,
ministry director, facilities manager].

Create a review form that includes:
- Self-assessment section (staff reflects on their
performance)
- Supervisor evaluation of key responsibilities (from
job description)
- Strengths and accomplishments this period
- Areas for growth or development
- Goal-setting for next review period
- Support or resources needed
- Overall performance rating (if required)

Review period: [annual, semi-annual, quarterly]
Church context: [size, personnel policies, evaluation
culture]
Tone: Constructive, growth-oriented, fair
```

What to Expect
AI will generate professional review templates with standard HR
sections. You'll need to customize with your church's specific job
descriptions, values, and evaluation criteria.

Follow-Up Prompts
"Add spiritual formation and character development
sections beyond just job performance."
Ministry positions require evaluating more than task completion –
character and spiritual health matter.
"Include 360-degree feedback components where staff
receive input from peers and those they supervise."
Comprehensive feedback provides fuller picture than just supervisor
perspective.
"Create development plan templates for staff needing
improvement in specific areas."
Reviews should lead to growth plans, not just documentation of
problems.

⚠ **Verify This**: Performance review processes must comply with
employment law and your denomination's personnel policies. Consult
HR or legal counsel before doing your first review.

🙂 **Your Discernment Matters**: Reviews should encourage growth, not
create fear. Balance accountability with pastoral care and grace.

🔒 **Privacy Reminder**: Performance reviews are confidential personnel
records. Store review files securely and share them only with appropriate
supervisors or denominational officials.

How to Use It
1. Use AI's review template and customize with your church's
   specific job expectations and values
2. Ask AI to create self-assessment guides helping staff prepare for
   reviews
3. Have AI develop training materials for supervisors conducting
   reviews fairly and constructively

Reflection Question
If this performance review process worked perfectly, how would it help
this staff member grow in both competence and spiritual maturity?

SECTION 7

CREATIVE & MEDIA

CHAPTER 90
Repurposing One Sermon into Content for Multiple Formats

Our Goal
Take one sermon and transform it into multiple content pieces: a blog post, social media graphics, an email devotional, and small group questions – all from a single source.

The Problem
You've invested hours preparing your sermon. It feels wasteful to preach it once and move on when that same content could serve your congregation (and reach new people) in multiple formats throughout the week. But creating all those versions manually is exhausting.

Why This Matters
Content repurposing enhances sustainable ministry communication. When you multiply one sermon into five formats, you extend its reach, reinforce its message, and meet people where they already are (email, social media, small groups) without creating new content from scratch every time.

The Prompt
```
I preached a sermon on [Scripture passage] titled
"[Sermon Title]." Here are the main points:
1. [Main point 1 with brief explanation]
2. [Main point 2 with brief explanation]
3. [Main point 3 with brief explanation]

Repurpose this sermon into the following formats:
1. A blog post (400-500 words) for our church website
2. Three social media posts (150 characters each)
with a call to action
3. A Monday morning email devotional (200 words) that
recaps the sermon
4. Five small group discussion questions
5. Three quote graphics (15-20 words each, suitable
for image overlays)

Audience: [Congregation description]
Tone: Warm, accessible, encouraging
```

What to Expect

AI will generate all five formats in a single response. The blog post will summarize key points, social posts will be punchy and shareable, the email will feel devotional, discussion questions will invite reflection, and quote graphics will highlight memorable phrases.

Follow-Up Prompts

"Make the social media posts more action-oriented with specific next steps."

This sharpens your call to action so people know exactly what to do, not just what to think.

"Rewrite the blog post in first-person as if I'm speaking directly to the reader."

Changing the voice can make content feel more intimate and conversational.

"Create three different versions of social post #2 so I can pick my favorite."

You can experiment with different approaches to see what your audience responds to best.

📖 **Scripture Check**: Ensure repurposed content accurately represents the Biblical text and sermon's theological message across all formats.

🙂 **Your Discernment Matters**: Different formats reach different audiences. Adjust tone and emphasis based on who's most likely to encounter each piece.

How to Use It

1. Use AI's repurposed content and personalize each format with your voice and local examples
2. Ask AI to create additional formats (Instagram stories, podcast clips, newsletter content)
3. Have AI develop a content calendar scheduling when each format gets published throughout the week

Reflection Question

Which format will reach people in your community who don't attend Sunday worship, and how might you use that format strategically for outreach rather than just member engagement?

CHAPTER 91
Creating Visual Graphics for Bulletins and Slides

Our Goal
Design simple, professional visual graphics for worship bulletins, presentation slides, and printed materials without being a graphic designer.

The Problem
You need visual elements for bulletins (announcements, sermon series branding, event promotions) and worship slides (sermon titles, Scripture verses, song lyrics), but you're not a designer and don't have budget for professional graphics. Generic clip art looks dated, but creating custom graphics takes skills and software you don't have.

Why This Matters
Visual design matters. Well-designed graphics communicate professionalism, enhance readability, and make information memorable. When AI helps you create graphic concepts and provides design guidance, you can produce quality visuals using free tools like Canva without design expertise.

The Prompt
```
I need visual graphics for [purpose: worship
bulletin, sermon series slides, event announcement,
social media graphic].

Create design concepts that include:
- Layout suggestions (text placement, visual
hierarchy)
- Color palette recommendations matching [theme:
Advent, Easter, sermon series topic, church branding]
- Typography guidance (font pairing, sizes,
readability)
- Image or icon suggestions
- Specific dimensions for [format: bulletin cover,
presentation slide, Instagram post, print flyer]

Content to include: [sermon title, Scripture verse,
event details, etc.]
Design style: [modern, traditional, minimalist, bold,
elegant]
Tools available: Canva, PowerPoint, free stock photos
```

What to Expect
AI will provide detailed design concepts with layout descriptions, color recommendations, and typography guidance. You'll still need to create the actual graphics using Canva or similar tools, but AI gives you the design strategy.

Follow-Up Prompts
"Suggest specific free stock photo sites where I could find images matching this concept."
AI can point you toward image resources that fit your design vision.
"Create three different layout options so I can choose which works best for our bulletin format."
Multiple design approaches help you pick what fits your space and aesthetic.
"Adjust the color palette to match our church's existing branding colors."
Design consistency across materials strengthens brand recognition.

☑ **Pro tip:** You can get great free photos and illustrations at Unsplash.com and Pixabay.com

☑ **Pro tip:** Creating graphics uses lots of "credits," so you may have to come back hours later to request your next revision or new image.

⚠ **Verify This**: Ensure all images used are properly licensed for church use. Not all "free" stock photos allow commercial or organizational use.

🙂 **Your Discernment Matters**: Visual design should enhance, not distract from, worship and message. Avoid overly trendy or cluttered graphics.

How to Use It
1. Use AI's output to help your volunteer designer or graphic artist better understand what you like and don't like
2. Ask AI to provide design templates for recurring needs (weekly bulletins, sermon series slides)
3. Have AI evaluate existing graphics and suggest improvements for clarity or visual appeal

Reflection Question
What visual style would best reflect your church's personality and values while remaining accessible to your congregation's demographics?

CHAPTER 92
Writing Image-Generation Prompts for Ministry Themes

Our Goal
Write effective prompts for AI image generators (like DALL-E, Midjourney, or Adobe Firefly) that create custom images for sermon series, events, or ministry themes.

The Problem
AI image generation tools can create custom graphics, but they require specific, detailed prompts to produce usable results. You need images for sermon series, event promotions, or social media, but don't know how to describe what you want in ways AI image generators understand.

Why This Matters
Custom-generated images can create unique visual branding for sermon series or events without stock photo costs or copyright concerns. When AI helps you write image-generation prompts, you can create original visual content even without photography or design skills.

The Prompt
```
I need to create an image-generation prompt for
[purpose: sermon series graphic, event poster, social
media image, bulletin cover] with the theme of
[theme: hope, redemption, community, journey, new
life, etc.].

Help me write an image-generation prompt that:
- Describes the visual concept clearly and
specifically
- Includes style preferences (photorealistic,
illustrated, watercolor, abstract, etc.)
- Specifies mood and atmosphere
- Includes technical details (composition, lighting,
color palette)
- Avoids elements that might be problematic or off-
brand

Target use: [worship slide, print materials, social
media]
Church context: [traditional, contemporary, diverse,
specific cultural considerations]
Desired feeling: [hopeful, contemplative, joyful,
peaceful, etc.]
```

What to Expect
AI will generate detailed image-generation prompts you can copy into AI art tools. The prompts will be specific enough to produce usable results while leaving room for creative interpretation.

Follow-Up Prompts
"Adjust this prompt to create a more abstract, symbolic representation rather than literal imagery."
Some themes work better as metaphor than realistic depiction.
"Add specific cultural or diversity elements ensuring the generated image represents our congregation's demographics."
AI image generators can perpetuate biases – intentional prompting creates more inclusive imagery.
"Create three variations of this prompt with different artistic styles so I can generate options."
Trying different styles helps you discover what resonates with your visual brand.

🌐 **Your Discernment Matters**: AI-generated images can include unintended religious symbolism, cultural insensitivity, or theological problems. Review carefully before use.

⚠ **Verify This**: Understand the licensing terms for AI-generated images. Some platforms restrict commercial or organizational use.

✝ **Theological Caution**: Visual representations of God, Jesus, or spiritual concepts can be theologically problematic. Approach sacred imagery with care and humility.

How to Use It
1. Use AI's prompts in image generation tools and iterate until you get usable results
2. Ask AI to help you refine prompts based on what the image generator actually produced
3. Have AI create prompt templates for recurring visual needs (sermon series, seasonal themes)

Reflection Question
What visual imagery would genuinely help your congregation engage with this theme without becoming cliché or theologically shallow?

CHAPTER 93
Developing Video Scripts for Online Worship or Outreach

Our Goal
Write compelling video scripts for online worship elements, ministry highlights, or outreach content that engage viewers and communicate clearly.

The Problem
Video is essential for online worship and social media outreach, but writing scripts that work on camera is different from writing sermons or print content. You need scripts that are conversational, visually engaging, and appropriate length for online attention spans.

Why This Matters
Video reaches people who won't read blog posts or attend in-person services. Well-scripted videos communicate clearly, stay on message, and respect viewers' time. When AI helps you write video scripts, you can create engaging content without video production expertise.

The Prompt
```
I'm creating a video script for [purpose: online
worship welcome, ministry highlight reel, stewardship
appeal, sermon teaser, event invitation].

Write a video script (60-90 seconds when spoken)
that:
- Opens with a compelling hook that grabs attention
in first 5 seconds
- Clearly communicates the main message
- Includes visual suggestions or B-roll descriptions
- Maintains conversational, camera-friendly language
- Ends with clear call to action

Video context: [platform: YouTube, Instagram,
Facebook, church website]
Speaker: [pastor, lay leader, testimonial, multiple
voices]
Tone: [warm, urgent, celebratory, contemplative]
```

What to Expect
AI will generate a complete video script with spoken dialogue and visual suggestions. The script will be written for verbal delivery, not reading, with natural speech patterns.

Follow-Up Prompts
"Add a personal testimony or story in the middle section to illustrate the message."
Stories make abstract messages concrete and create emotional connection.
"Shorten this to 30 seconds for Instagram Reels or TikTok."
Different platforms require different video lengths and pacing.
"Include specific visual transitions or shot suggestions for the videographer."
Detailed visual direction helps even amateur videographers produce quality content.

🔒 **Privacy Reminder**: Get written permission from anyone featured in video content, especially children. Not everyone wants to be on camera representing your church.

👁 **Your Discernment Matters**: Video scripts should sound natural when spoken, not written. Read scripts aloud and revise awkward phrases.

How to Use It
1. Use AI's script as your foundation and adjust language to match the speaker's natural voice
2. Ask AI to create multiple script versions for A/B testing which messages resonate
3. Have AI develop shot lists and production notes accompanying the script

Reflection Question
What one message, if communicated effectively in 30, 60, or 90 seconds, would most help someone understand what your church is about or why this ministry matters?

CHAPTER 94
Creating Podcast Episode Scripts and Show Notes

Our Goal
Write podcast episode scripts and show notes that create engaging audio content and help listeners find and understand episodes.

The Problem
Podcasting extends your teaching ministry beyond Sunday, but creating episode content requires different skills than sermon writing. You need conversational scripts, compelling episode descriptions, and organized show notes. Producing consistent podcast content alongside all your other responsibilities may feel overwhelming.

Why This Matters
Podcasts meet people in their daily routines – commutes, workouts, chores. Well-scripted episodes feel professional and stay on message while maintaining conversational warmth. Good show notes help people find episodes and provide resources for going deeper. When AI helps you create podcast content, you can launch or sustain a podcast without doubling your workload.

The Prompt
```
I'm creating a podcast episode for [show name] on the
topic of [theme or Scripture passage].

Create podcast content that includes:

**Episode script outline (20-30 minutes):**
- Opening hook and episode intro
- Main content sections with talking points
- Stories, examples, or illustrations
- Transition phrases between sections
- Closing summary and call to action

**Show notes:**
- Episode title and description (150 words)
- Key takeaways (3-5 bullet points)
- Scripture references and resources mentioned
- Listener action steps
- Social media teaser (50 words)

Podcast style: [solo teaching, interview,
conversational co-hosts]
```

```
Audience: [church members, spiritual seekers, topic-
specific]
Tone: Conversational, accessible, engaging
```

What to Expect
AI will generate both episode script outline and complete show notes. The script will be conversational rather than formal, suitable for speaking naturally rather than reading verbatim.

Follow-Up Prompts
```
"Add interview questions if I'm bringing a guest to
discuss this topic."
```
Interview episodes need different structure than solo teaching episodes.
```
"Create social media promotional posts for this
episode with compelling hooks."
```
Podcast promotion requires different messaging than the episode itself.
```
"Suggest relevant books, articles, or resources to
include in show notes for listeners who want to go
deeper."
```
Show notes add value by pointing listeners to additional resources.

📖 **Scripture Check**: Ensure all Biblical references and theological content in the script are accurate and contextually sound.

🧑 **Your Discernment Matters**: Podcast scripts are guides, not verbatim reading. Speak naturally and follow conversational flow rather than rigid adherence to script.

How to Use It
1. Use AI's script outline as your episode framework but speak naturally rather than reading word-for-word
2. Ask AI to create episode series outlines for multi-week podcast seasons
3. Have AI develop listener questions or feedback prompts for future episodes

Reflection Question
What aspect of your teaching or ministry would benefit from the longer, more conversational format podcasting allows compared to Sunday sermons?

CHAPTER 95
Writing Captions and Alt-Text for Accessibility

Our Goal
Write descriptive captions and alt-text for images that make visual content accessible to people using screen readers or needing additional context.

The Problem
Accessibility matters, but you may not know how to write effective alt-text or captions. People who are visually impaired use screen readers that read alt-text aloud. Others benefit from captions explaining visual content. Writing these descriptions for every image you post may feel like one more task on an endless list.

Why This Matters
Accessible content ensures everyone can participate in your church's digital ministry regardless of disability. Alt-text and captions demonstrate care for all people and often improve everyone's experience. When AI helps you write these descriptions, you can make content accessible without extensive training in accessibility standards.

The Prompt
```
I need to write alt-text and captions for [image
description: photo of worship service, ministry
event, Scripture graphic, illustrated quote, etc.].

Create accessibility content that includes:
- Alt-text description (concise, descriptive,
conveying essential information for screen readers)
- Extended caption (providing context, emotion, or
details beyond what's visible)
- Relevant hashtags if appropriate for social media

Image context: [where this will be used: website,
social media, bulletin, email]
Key information to convey: [what's important about
this image]
Tone: Descriptive, clear, helpful
```

☑ **Pro tip:** If you attach the photos that need captions to your prompt, your AI will generate incredibly specific results.

What to Expect

AI will generate both concise alt-text suitable for screen readers and fuller captions providing context. The descriptions will focus on relevant details without unnecessary decoration.

Follow-Up Prompts

```
"Make the alt-text more concise – screen reader users
don't need every visual detail."
```
Good alt-text describes what matters, not every element visible in the image.
```
"Add emotional context to the caption – describe the
feeling or atmosphere of this moment."
```
Extended captions can convey mood and significance beyond literal description.
```
"Create alt-text that includes the text visible in
this graphic or image."
```
Text embedded in images must be described in alt-text for accessibility.

🔒 **Privacy Reminder**: When captioning photos of people, consider privacy. Some individuals may not want to be named or identified in alt-text.

😊 **Your Discernment Matters**: Alt-text should describe what's relevant to understanding the image's purpose, not provide exhaustive visual inventory.

How to Use It

1. Use AI's alt-text as your starting point and edit for conciseness and relevance
2. Ask AI to create alt-text templates for recurring image types (bulletin covers, sermon graphics, event photos)
3. Have AI develop accessibility checklists ensuring all digital content includes proper descriptions

Reflection Question

If someone couldn't see this image, what information would they need to understand why it matters and what it contributes to the content?

CHAPTER 96
Developing Shareable Infographics for Complex Concepts

Our Goal
Create shareable infographics that visually explain abstract concepts like discipleship stages, decision trees, theological ideas, or ministry processes in accessible, engaging formats.

The Problem
Some information works better visually than in paragraphs. You want to create infographics that are both informative and shareable on social media, but graphic design and information architecture aren't your skills.

Why This Matters
Infographics make complex information accessible and memorable. They're highly shareable on social media, useful in classes and presentations, and help visual learners engage with content. When AI helps you design infographic concepts, you can create educational visuals without professional design expertise.

The Prompt
```
I want to create an infographic explaining [topic:
discipleship pathway, theological concept, church
governance, spiritual gifts, prayer practices, etc.].

Design an infographic concept that includes:
- Main visual structure (flowchart, circular diagram,
timeline, comparison chart, pyramid, etc.)
- Key information organized into clear sections
- Text content for each section (brief, scannable)
- Visual hierarchy (what's most prominent)
- Color coding or visual cues grouping related
information
- Suggested icons or simple illustrations

Audience: [church members, seekers, social media
followers]
Complexity: [simple overview, detailed explanation,
quick reference]
Platform: [Instagram, bulletin insert, presentation
slide, website]
Dimensions: [square for Instagram, vertical for
Pinterest, landscape for slides]
```

What to Expect
AI will provide detailed infographic design concepts with content organized visually, structural suggestions, and text for each section. You'll still create the actual graphic using Canva or similar tools, but AI provides the information architecture.

Follow-Up Prompts
"Simplify this to fit on a single Instagram post – it's too complex for social media."
Different platforms require different levels of detail and visual complexity.
"Add specific Scripture references supporting each section of this theology explainer."
Biblical grounding makes theological infographics more credible and useful.
"Create a series of three related infographics that build on each other."
Complex topics sometimes need multiple infographics rather than cramming everything into one.

📖 **Scripture Check**: Theological infographics must accurately represent Biblical teaching. Oversimplification can create theological misconceptions.

✝ **Theological Caution**: Visual representations of complex theology risk reductionism. Ensure infographics invite deeper learning rather than replacing it.

🙂 **Your Discernment Matters**: Not all information works well in infographic format. Some concepts require narrative or conversation, not diagrams.

How to Use It
1. Use AI's infographic concept and create the actual visual in Canva or PowerPoint
2. Ask AI to provide different structural approaches (flowchart vs. circular vs. hierarchical) and pick what fits the content
3. Have AI develop supporting explanatory text that could accompany the infographic

Reflection Question
What concept or process in your church would people understand better with a visual diagram than with verbal explanation?

CHAPTER 97
Creating Worship Lyric Slides with Formatting

Our Goal
Format song lyrics for worship slides that are readable, visually appealing, and support congregational singing.

The Problem
Projecting lyrics seems simple, but poor formatting undermines worship. Text that's too small, inconsistent line breaks, cluttered slides, or awkward phrasing makes singing difficult. You want slides that enhance worship, not distract from it, but don't have design training or know lyric formatting best practices.

Why This Matters
Good slide design and well-formatted lyrics help people sing confidently and focus on worship rather than squinting at screens. AI can help you format lyrics and create readable slides without hours of trial-and-error design work.

The Prompt
```
I need to format lyrics for [song title] for worship
projection slides.

Here are the complete lyrics:
[Paste full song lyrics or attach PDF of sheet music]

Create slide formatting that:
- Breaks lyrics into readable slide segments (2-4
lines per slide)
- Maintains natural phrasing and breathing points
- Indicates repeats, tags, or structural elements
- Suggests visual hierarchy (chorus vs. verse
formatting)
- Recommends font size, line spacing, and slide
background
- Notes any special considerations (bridge,
instrumental breaks)

Sanctuary context: [screen size and distance from
congregation]
Technical setup: [ProPresenter, PowerPoint, other
software]
```

```
Song style: [traditional hymn, contemporary worship,
responsive reading]
```

What to Expect
AI will provide complete slide-by-slide lyric formatting with line breaks, structural notes, and design recommendations. You'll copy this formatting into your presentation software.

Follow-Up Prompts
```
"Adjust line breaks to ensure phrases don't awkwardly
split across slides."
```
Musical phrasing should guide slide breaks, not just character count.
```
"Add visual cues showing when to repeat the chorus or
bridge."
```
Clear repeat indicators help worship leaders and musicians stay coordinated.
```
"Suggest background colors or images that won't
compete with lyric readability."
```
Backgrounds should enhance, not obscure, text readability.

⚠ **Verify This**: Ensure you have proper licensing (CCLI, OneLicense, etc.) for all songs you project. Copyright compliance is required and honors the efforts of the original writers and composers.

🙂 **Your Discernment Matters**: Readability trumps aesthetics. If decorative elements compromise legibility, simplify.

How to Use It
1. Apply AI's lyric formatting suggestions to the lyrics in your presentation software (ProPresenter, PowerPoint, etc.)
2. Ask AI to create formatting templates for different song types (hymns, contemporary, responsive readings)
3. Have AI develop slide design guidelines ensuring consistency across worship services

Reflection Question
Can someone in the back row with moderate vision challenges easily read and follow along with these lyrics, or do they need to guess?

CHAPTER 98
Writing Scripts for Children's Plays

Our Goal

Write engaging, age-appropriate scripts for children's Christmas pageants, VBS performances, or church anniversary celebrations that involve kids in meaningful ways.

The Problem

Children's performances are beloved church traditions, but writing scripts that are theologically sound, developmentally appropriate, funny without being silly, and actually engaging for both performers and audience is challenging. You want children to learn while having fun, not just recite lines they don't understand.

Why This Matters

Children's performances create memories, build confidence, and teach Biblical stories in embodied ways. When AI helps you write scripts, you can create quality performances without playwriting expertise.

The Prompt

```
I'm writing a children's play script for [occasion:
Christmas pageant, Easter presentation, VBS closing
program, church anniversary] for children ages [age
range].

The key Bible verses are: [insert verses]
The key themes for the play are: [insert themes]

Create a script (10-15 minutes) that includes:
- Biblical story or church theme presented age-
appropriately
- 8-12 speaking roles (with options for non-speaking
roles)
- Simple, memorable dialogue kids can learn
- Stage directions and movement
- Optional songs or music cues
- Narrator sections connecting scenes
- Ending that reinforces the Biblical message

Performance context: [sanctuary, fellowship hall,
outdoor]
Audience: [congregation, families, community]
Tone: Joyful, reverent, accessible to young children
```

```
Biblical accuracy: Faithful to the core concepts in
the Scripture reading
```

What to Expect
AI will generate a complete play script with dialogue, stage directions, and role assignments. You'll need to adjust based on your actual children's abilities, available costumes/props, and performance space.

Follow-Up Prompts
```
"Add more non-speaking roles so we can include
younger children who can't memorize lines."
```
Inclusive scripts let children of all ages participate meaningfully.
```
"Create a simplified narrator version that tells the
story if we don't have time for full dialogue
rehearsal."
```
Backup simpler versions reduce stress when rehearsal time is limited.
```
"Rewrite this Biblical story; set it in a 21st-
century suburban context using contemporary teenage
language and situations."
```
This helps you modernize the story to make it more relatable and accessible to today's children while preserving the Biblical truth.

 Your Discernment Matters: Scripts should be age-appropriate for both performers and audience. Avoid theology too complex for children or humor inappropriate for worship.

 Theological Caution: Christmas and Easter scripts often add sentimental or unbiblical elements. Stay faithful to Gospel accounts **or** ensure the audience understands the play is an artistic interpretation based on the Scripture reading.

How to Use It
1. Use AI's script and adjust dialogue based on your children's actual speaking abilities
2. Ask AI to create costume and prop lists keeping production simple and affordable
3. Have AI develop rehearsal schedules and parent communication about performance expectations

Reflection Question
What do you most want children and families to learn or experience through this performance beyond just putting on a cute show?

CHAPTER 99
Developing Email Signature Designs with Ministry Info

Our Goal
Create professional email signatures for staff and key volunteers that include essential contact information and ministry branding.

The Problem
Email signatures should look professional, include necessary contact details, and reflect your church's branding. But most people don't know how to format email signatures that work across different email clients without looking broken. Inconsistent signatures across your team look unprofessional.

Why This Matters
Professional email signatures build credibility, make it easy for people to contact your church, and extend your brand into every email your team sends. When AI helps you design signatures, you can create consistent, professional templates for your entire staff and volunteer leadership.

The Prompt
```
I'm creating an email signature template for [role:
senior pastor, staff member, ministry leader, church
office].

Design an email signature that includes:
- Name and title/role
- Church name
- Contact information (phone, email, office hours if
relevant)
- Church address and website
- Social media links (optional)
- Church or denominational logo [attach logo file]
- Mobile-friendly formatting

Church context: [formal vs. casual, traditional vs.
contemporary]
Email platform: [Gmail, Outlook, other]
Design style: [minimal, branded, professional]
```

What to Expect
AI will provide email signature design concepts with layout suggestions and formatting guidance. You'll need to create the actual signature in your email platform or using an email signature generator.

Follow-Up Prompts
"Create a simpler version without graphics that will display correctly in all email clients."
Some email platforms don't handle images well – text-only versions ensure consistency.
"Add event invitation or seasonal greeting line."
Variable elements can highlight current sermon series, events, or seasonal greetings.
"Suggest different signature versions for staff vs. volunteers vs. pastors."
Different roles may need different levels of contact information and formality.

⚠ **Verify This**: Test email signatures across different email platforms (Gmail, Outlook, mobile) to ensure they are easy to read.

🧠 **Your Discernment Matters**: Email signatures should be informative without being cluttered. Include only essential information.

🔒 **Privacy Reminder**: Consider whether personal cell phone numbers should be in email signatures or just church office numbers.

How to Use It
1. Use AI's signature format to create actual signatures using email signature generators or HTML
2. Ask AI to create a "Church Email Signature Style Guide" to ensure consistency across your team
3. Have AI develop signature update schedules for seasonal messaging or staff changes

Reflection Question
What impression does your current email signature create about your church's professionalism and accessibility, and does it need updating?

SECTION 8

RESEARCH & PROFESSIONAL DEVELOPMENT

CHAPTER 100
Summarizing Theological Articles and Commentaries

Our Goal
Quickly extract key insights from theological articles, Biblical commentaries, or scholarly papers without reading every word.

The Problem
Sermon prep and continuing education require reading theological scholarship, but you don't have time to read 10-page journal articles or multi-volume commentaries cover-to-cover. You need the main points, Biblical insights, and relevant applications.

Why This Matters
Theological reading strengthens your preaching and deepens your understanding, but time constraints are real. When AI helps you summarize scholarly sources, you can engage with continuous learning and growth without sacrificing other pastoral responsibilities.

The Prompt
```
I'm reading [article title, author, or commentary on
Scripture passage] for sermon preparation or
continuing education.

[If possible, attach a PDF of the source material.]

Summarize this source focusing on:
- Main thesis or argument
- Key Biblical or theological insights
- Important quotes or memorable phrases
- How this connects to [Scripture passage I'm
preaching or studying]
- Potential sermon applications
- Critical questions or limitations of this
perspective

Source type: [journal article, commentary section,
theological essay]
Length: 200-300 words
Focus: Practical insights for preaching and teaching
```

What to Expect

AI will generate a concise summary highlighting the most relevant content for your ministry needs. You'll get the essence without reading the entire source, though you should still verify key claims.

Follow-Up Prompts

```
"What are the strongest critiques of this author's
argument?"
```
Understanding both support and opposition helps you evaluate theological claims critically.
```
"How does this perspective differ from [another
theological tradition or scholar]?"
```
Comparing perspectives helps you understand debates and choose wisely among options.
```
"Extract specific quotes I could cite in a sermon or
class."
```
Finding quotable moments saves time searching through long documents.

📖 **Scripture Check**: Even when summarizing trusted sources, verify all Biblical references and theological claims. Summaries can lose important nuances.

✝ **Theological Caution**: AI summaries may miss subtleties or misrepresent complex arguments. Read relevant portions of original sources when making significant theological decisions.

⚠ **Verify This**: Always cite sources properly. Summarizing isn't plagiarizing, but you must attribute ideas to their authors.

How to Use It

1. Use AI summaries to decide which sources deserve your full reading time
2. Ask AI to summarize multiple perspectives on a topic for comprehensive understanding
3. Have AI create annotated bibliographies organizing sources by theme or perspective

Reflection Question

Which theological insight from this summary most challenges or enriches your current thinking about this Scripture or topic?

CHAPTER 101
Researching Complex Theological Topics Quickly

Our Goal
Research complex theological topics efficiently, understanding major positions and arguments without months of study.

The Problem
Theological questions arise constantly – from sermon prep, pastoral conversations, or your own curiosity. You need to understand debates about atonement theories, eschatology, or sacramental theology, but don't have seminary course time. You need reliable information quickly.

Why This Matters
Theological literacy strengthens your preaching, pastoral care, and leadership. Understanding multiple perspectives helps you navigate congregational diversity and teach with nuance. When AI helps you research theology quickly, you can engage complex topics without extensive library research.

The Prompt
```
I'm researching the theological topic of [topic:
atonement theories, predestination, baptism theology,
women in ministry, LGBTQ inclusion, etc.].

Provide a research overview that includes:
- Definition and why this matters theologically
- Major theological positions (2-4 main perspectives)
- Key Biblical passages each position emphasizes
- Historical development of the debate
- Where my tradition ([denomination]) typically
stands
- Resources for deeper study

Audience: Pastor needing working knowledge
Theological framework: Accessible overview, not
academic detail
Length: 400-500 words
```

What to Expect
AI will generate a comprehensive overview with multiple perspectives, Biblical grounding, and denominational context. You'll get working knowledge without reading multiple books.

Follow-Up Prompts
"Add specific counter-arguments against each position
so I understand the full debate."
Understanding objections helps you evaluate positions critically and
engage diverse views charitably.
"Explain this at a level appropriate for confirmation
students or adult education classes."
You can adjust complexity for different teaching contexts.
"Suggest 3-5 trusted books or articles for each major
position if I want to study further."
Research overviews should point toward deeper learning, not replace it.

✅ **Pro tip:** As of March 2026, Perplexity.ai excels at academic research
and providing detailed bibliographies of sources it's found.

📖 **Scripture Check**: Theological research must be grounded in
Scripture. Verify that AI's summary accurately represents Biblical texts
supporting each position.

✝️ **Theological Caution**: AI provides overviews, not authoritative
theological teaching. Consult trusted scholars and denominational
resources for serious theological decisions.

🤔 **Your Discernment Matters**: On contested issues, AI may present
positions as equally valid when your tradition has clear convictions.
Know your theological commitments.

How to Use It
1. Use AI's overview as starting point and read recommended
   sources for positions you're considering
2. Ask AI to research multiple related topics to understand
   theological frameworks comprehensively
3. Have AI create study guides or discussion materials based on
   your research

Reflection Question
Which theological position in this overview is most challenging to your
current thinking, and what would it take for you to seriously consider it?

CHAPTER 102
Analyzing Congregational Surveys

Our Goal
Analyze congregational survey data to identify trends, concerns, and opportunities for ministry improvement.

The Problem
You've collected survey responses about worship satisfaction, small group participation, or strategic planning priorities, but now you're staring at hundreds of responses without knowing how to make sense of the data. You need to identify patterns, summarize feedback, and translate data into actionable ministry insights.

Why This Matters
Congregational surveys provide valuable input for decision-making, but raw data is overwhelming. Clear analysis helps leadership understand what people actually think and need. When AI helps you analyze surveys, you can extract meaningful insights without statistical expertise.

The Prompt
```
I conducted a congregational survey with [number]
responses about [topic: worship preferences, small
group satisfaction, facility needs, strategic
priorities, pastoral care effectiveness].

Here are the survey results:
[Attach a PDF of the surveys or paste summary data,
response counts, or key questions with answer
distributions]

Analyze this data and provide:
- Key themes and trends
- Areas of strong consensus or significant division
- Surprising or notable findings
- Demographic patterns if relevant (age, tenure,
involvement level)
- Recommendations for leadership action
- Follow-up questions we should explore

Church context: [size, demographics, current
challenges]
Survey purpose: [inform strategic planning, evaluate
ministry, gather input on decision]
```

What to Expect
AI will identify patterns, summarize themes, and suggest implications for ministry. You'll get data interpretation without needing statistical software or analysis training.

Follow-Up Prompts
"Compare responses between long-time members and newer members to see if perspectives differ."
Demographic analysis reveals whether different groups have different needs or concerns.
"Identify the top 3 actionable priorities leadership should address based on this data."
Moving from analysis to action requires prioritizing what matters most.
"Draft a summary report presenting these findings to the congregation or church board."
Survey results need clear communication to stakeholders who provided input.

⚠ **Verify This**: Ensure AI's analysis accurately reflects your actual survey data. Don't let AI invent findings not present in responses.

🙂 **Your Discernment Matters**: Survey data shows what people think, not necessarily what's best for the church. Prayerful discernment must guide decisions, not just popular opinion.

How to Use It
1. Use AI to identify themes in open-ended responses where patterns aren't immediately obvious
2. Ask AI to create visual charts or graphs representing survey data for presentations
3. Have AI draft communication materials sharing survey results with the congregation

Reflection Question
What does this survey data reveal about gaps between what your church currently offers and what your congregation actually needs or wants?

CHAPTER 103
Tracking Attendance and Giving Trends with Data Analysis

Our Goal
Analyze attendance and giving data to understand trends, identify concerns, and make informed decisions about ministry and budget.

The Problem
Your church tracks attendance and giving, but the numbers sit in spreadsheets without meaningful analysis. You need to understand trends – are you growing or declining? Are seasonal patterns normal or concerning? Does giving align with attendance? You need insights, not just data.

Why This Matters
Data-informed leadership makes better decisions. Understanding attendance and giving trends helps you plan staffing, create balanced budgets, and identify ministry opportunities or concerns early. When AI helps you analyze data, you can understand patterns without becoming a data analyst.

The Prompt
```
I'm analyzing church data to understand trends and
inform planning.

Here's the data:
[Provide: monthly attendance numbers for past 1-3
years, giving totals, comparison to previous years,
seasonal patterns, etc. Attach data files if
possible.]

Analyze this data and provide:
- Overall trends (growth, decline, stability)
- Seasonal patterns and what's normal vs. concerning
- Relationship between attendance and giving
- Comparison to previous years
- Potential explanations for significant changes
- Recommendations for leadership response

Church context: [size, location, recent changes or
events]
Time period: [monthly data for X years]
```

What to Expect
AI will identify trends, seasonal patterns, and correlations between attendance and giving. You'll get interpretation helping you understand what the numbers mean for ministry planning.

Follow-Up Prompts
```
"Project attendance and giving trends forward 3 years
based on current trajectory."
```
Forecasting helps with budget planning and identifies when intervention might be needed.
```
"Identify which months show unusual patterns and
suggest possible explanations."
```
Understanding anomalies helps you determine whether changes are random fluctuations or meaningful trends.
```
"Compare our trends to typical patterns for churches
our size and demographic context."
```
Context helps determine whether your trends are concerning or normal for your situation.

⚠ **Verify This**: Data analysis is only as good as the data quality. Ensure attendance and giving records are accurate before analyzing.

🙂 **Your Discernment Matters**: Numbers tell part of the story, not the whole story. Faithfulness to mission matters more than numerical growth.

How to Use It
1. Use AI's trend analysis to inform budget conversations and staffing decisions
2. Ask AI to create visual presentations of data for board meetings or congregational reports
3. Have AI identify early warning signs of financial or attendance concerns requiring attention

Reflection Question
What do these trends reveal about your church's health and trajectory, and what leadership decisions might they inform?

CHAPTER 104
Creating Continuing Education Plans and Learning Goals

Our Goal
Design personal continuing education plans with specific learning goals, resources, and accountability that keep you growing as a pastor and leader.

The Problem
You know you should keep learning, but continuing education may feel haphazard – random conferences, occasional reading, scattered online courses. You need a strategic plan for growth that addresses your actual learning needs and ministry challenges, but designing comprehensive professional development plans takes time and self-awareness.

Why This Matters
Intentional learning prevents stagnation and equips you for evolving ministry challenges. Strategic continuing education addresses weaknesses, deepens strengths, and keeps you intellectually and spiritually engaged. When AI helps you create learning plans, you can design growth strategies without extensive planning overhead.

The Prompt
```
I'm creating a continuing education plan for the next
[6 months/1 year/3 years] focusing on [areas:
preaching, leadership, pastoral care, theology,
administration, spiritual formation].

Create a learning plan that includes:
- Specific, measurable learning goals in each area
- Recommended resources (books, courses, conferences,
mentors)
- Timeline and milestones
- How learning will apply to my ministry context
- Accountability structure
- Budget estimate for paid resources

Current context: [years in ministry, current role,
areas needing growth]
Learning style: [reading, courses, mentoring,
experiential]
Time available: [hours per week for professional
development]
Budget: [available for continuing education]
```

What to Expect
AI will generate a structured learning plan with specific goals, resource recommendations, and realistic timelines. You'll get a roadmap for growth tailored to your needs and constraints.

Follow-Up Prompts
```
"Prioritize these goals – what should I focus on
first based on my current ministry challenges?"
```
Not all learning is equally urgent. Strategic sequencing helps you address pressing needs first.
```
"Suggest free or low-cost alternatives to the
expensive conferences or courses recommended."
```
Quality learning doesn't always require large budgets.
```
"Add specific accountability measures – how will I
know if I'm actually learning and growing?"
```
Learning plans need assessment, not just activity checklists.

⚠ **Verify This**: Recommended resources should align with your theological tradition and ministry context. Not all learning opportunities are equally valuable.

How to Use It
1. Use AI's learning plan and adjust based on actual time and budget realities
2. Ask AI to create quarterly check-in questions assessing progress and adjusting goals
3. Have AI develop learning cohorts or reading groups with colleagues pursuing similar growth

Reflection Question
What skill, knowledge area, or spiritual capacity would most strengthen your ministry if you invested serious learning time in it over the next year?

CHAPTER 105
Researching Denominational Polity and Governance Questions

Our Goal
Quickly research denominational policies, governance procedures, and polity questions without spending hours in church law manuals.

The Problem
Denominational governance is complex. You need to know proper procedures for trustee decisions, pastoral transitions, property matters, or conflict resolution, but *Book of Discipline* paragraphs can be dense or confusing. You need quick answers to procedural questions.

Why This Matters
Following proper denominational procedures protects your church legally, maintains accountability, and honors your covenant with the broader church. Mistakes in governance can be costly and time-consuming to fix. When AI helps you research polity, you can navigate denominational requirements without becoming a constitutional expert.

The Prompt
```
I have a denominational polity question about [topic:
pastoral appointment process, property decisions,
trustee authority, membership transfer, clergy
discipline, church closure, etc.] in the
[denomination].

Research this question and provide:
- Relevant sections from our polity documents (Book
of Discipline, constitution, bylaws)
- Step-by-step procedures to follow
- Who has authority to make this decision
- Common mistakes or pitfalls to avoid
- Where to get additional authoritative guidance

Denomination: [UMC, Presbyterian, Baptist, Lutheran,
etc.]
Context: [specific situation if relevant]
Urgency: [routine question vs. immediate need]
```

What to Expect
AI will summarize relevant polity provisions and procedures. For denominational-specific questions (especially the UMC), use AI chatbots like "Wes" (Chapter 11) for specific denominational guidance.

Follow-Up Prompts
```
"What are the consequences if we don't follow this
procedure correctly?"
```
Understanding stakes helps you know when to be meticulous versus when informal processes suffice.
```
"Who in our denominational structure (District
Superintendent, conference staff) should I consult
about this?"
```
Some questions require official denominational guidance beyond research.
```
"Compare how our denomination handles this versus how
other traditions approach it."
```
Understanding different approaches can inform your advocacy for policy changes.

⚠ **Verify This**: AI provides general guidance, not authoritative legal or denominational interpretation. Always verify polity questions with official denominational resources or leaders.

🙂 **Your Discernment Matters**: Polity exists to serve mission, not obstruct it. Follow procedures while advocating for faithful ministry.

How to Use It
1. Use AI for initial research, then verify with denominational officials before major decisions
2. Ask AI to create governance checklists for common procedures your church handles regularly
3. Have AI develop leadership training materials explaining denominational polity to board members

Reflection Question
If this procedure feels burdensome or inappropriate for your context, what's the faithful way to navigate that tension – compliance, advocacy for change, or creative adaptation?

CHAPTER 106
Developing Reading Lists for Personal or Group Study

Our Goal
Create curated reading lists for personal growth or group study that address specific topics, balance perspectives, and include accessible resources.

The Problem
You want to read more about preaching, leadership, theology, or spiritual formation, but don't know which books to choose. Or you're starting a clergy reading group or adult education series and need quality book recommendations. Creating reading lists requires knowledge of available resources and ability to evaluate quality.

Why This Matters
Good reading shapes faithful ministry. Well-curated lists save time, ensure quality, and create coherent learning journeys. When AI helps you develop reading lists, you can discover resources you wouldn't find otherwise and design strategic reading plans.

The Prompt
```
I'm creating a reading list on [topic: preaching,
pastoral care, church leadership, spiritual
formation, social justice, specific theology] for
[audience: personal reading, clergy group, adult
education class, confirmation students].

Develop a reading list that includes:
- 5-10 books spanning introductory to advanced
- Brief description of each book's contribution
- Balance of theological perspectives where relevant
- Mixture of classic and contemporary resources
- Estimated reading timeline
- Discussion guide suggestions if for group study

Reading context: [available time, reading level,
budget considerations]
Theological tradition: [preferred perspective or
desire for diverse views]
Purpose: [personal growth, group study, teaching
preparation]
```

What to Expect
AI will generate a curated reading list with book descriptions and reading progression. You'll get strategic recommendations beyond bestseller lists or random suggestions.

Follow-Up Prompts
```
"Identify which books are essential starting points
versus advanced reading for later."
```
Sequencing helps readers build knowledge progressively rather than jumping into deep water.
```
"Add books representing perspectives I might disagree
with for balanced understanding."
```
Reading widely strengthens thinking even when you don't adopt every view.
```
"Here are some authors I particularly value: [list
names]. Adjust your reading list to include books by
these authors or similar writers."
```
This helps you shape the reading list around theological voices and writing styles you already trust and appreciate.

✝ **Theological Caution**: Verify that recommended books align with your tradition's core convictions, especially for group study where you're implicitly endorsing content.

🙂 **Your Discernment Matters**: Not every popular or well-reviewed book is worth your limited reading time. Choose what brings you joy.

How to Use It
1. Use AI's reading list and adjust based on books you've already read or specific interests
2. Ask AI to create discussion guides or reflection questions for group reading
3. Have AI develop reading schedules showing how to complete the list over specific timeframes

Reflection Question
What topic or question, if you studied it deeply through quality reading, would most strengthen your ministry in the next season?

CHAPTER 107
Summarizing Books for Sermon Prep or Teaching

Our Goal
Quickly extract key insights from books for newsletter articles, sermon preparation, teaching, or ministry application without reading every page.

The Problem
Books offer deep insights, but you don't have time to read entire volumes for every sermon or class. You need to understand a book's main arguments, memorable quotes, and practical applications efficiently.

Why This Matters
Summaries help you decide which books deserve full reading and extract value from others quickly. When AI helps you summarize books, you can benefit from broader reading within your limited time demands.

The Prompt
```
I'm reading [book title and author] for [purpose:
newsletter article, sermon prep on specific topic,
teaching preparation, understanding an issue,
personal growth].

Summarize this book focusing on:
- Main thesis and key arguments
- Chapter-by-chapter breakdown of major ideas
- Most relevant insights for [my specific purpose]
- Memorable quotes suitable for sermons or teaching
- Practical applications to ministry
- Critical evaluation (strengths and limitations)

Reading purpose: [sermon on specific passage, class
on topic, general knowledge]
Depth needed: [quick overview vs. detailed summary]
Length: 300-400 words
```

What to Expect
AI will provide comprehensive summary with chapter breakdowns, key quotes, and ministry applications. You'll understand the book's essence without full reading, though summaries can't replace deep engagement.

Follow-Up Prompts
```
"Which chapters are most relevant to my sermon on
[Scripture passage or topic]?"
```
Targeted reading saves time when you only need portions of a book.
```
"What are the strongest critiques of this book's
arguments?"
```
Understanding objections helps you use ideas critically and anticipate questions.
```
"Extract 3-5 quotes I could cite in sermons or
teaching."
```
Finding quotable material quickly enhances your teaching without extensive searching.

⚠ **Verify This**: Read relevant parts of the original sources to confirm the accuracy of your summary before engaging in major theological decisions or contested issues.

✝ **Theological Caution**: AI summaries can oversimplify complex theology, miss Biblical nuances, or misrepresent theological arguments. Ensure theological claims are connected to specific Scriptures and are in line with your denominational teachings.

How to Use It
1. Use summaries to decide which books deserve your full reading time
2. Ask AI to summarize multiple books on the same topic for comparative understanding
3. Have AI create teaching outlines or discussion guides based on book summaries

Reflection Question
Which insight from this book summary most challenges your current thinking or practice, and what would change if you took it seriously?

CHAPTER 108
Analyzing Engagement Data

Our Goal
Analyze engagement data from sermons, social media, or newsletters to understand what content resonates with your audience and why.

The Problem
You publish sermons online, post to social media, and send newsletters, but often don't know what's actually connecting with people. Which sermon topics get the most views? Which social posts get shared? What email subject lines get opened? You have analytics data but no idea what it means.

Why This Matters
Understanding what resonates helps you communicate more effectively. Data reveals gaps between what you think matters and what actually connects with people. When AI helps you analyze engagement, you can make informed decisions about content strategy without becoming a data scientist.

The Prompt
```
I'm analyzing engagement data to understand what
content resonates with our audience.

Here's the data:
[Provide: sermon view counts by topic, social media
engagement metrics, email open rates, website traffic
by page, etc.]

Analyze this data and identify:
- Content themes or topics that consistently perform
well
- Formats or platforms that generate most engagement
- Patterns in timing or posting schedules
- Gaps between what we produce and what people engage
with
- Recommendations for content strategy
- Potential explanations for surprising results

Church context: [size, demographics, ministry focus]
Platforms: [YouTube, Facebook, Instagram, email,
website]
Time period: [data from past 3-6-12 months]
```

What to Expect
AI will identify patterns in engagement data, highlight what's working, and suggest strategic adjustments. You'll understand what content connects and why.

Follow-Up Prompts
"Compare engagement across different demographics if we have that data (age groups, member vs. visitor, etc.)."
Different audiences engage differently – segmented analysis reveals distinct preferences.
"Suggest content topics or formats we should try based on what's resonating."
Analysis should lead to action, not just observation.
"Identify which low-engagement content might still be important to continue despite metrics."
Not everything valuable is popular – some content serves mission more than metrics.

🧐 **Your Discernment Matters**: High engagement doesn't always mean high value. Don't let metrics drive all content decisions. Some important messages won't be popular.

⚠ **Verify This**: Ensure data is being measured accurately. Vanity metrics (likes, views) may not reflect genuine impact or spiritual formation.

How to Use It
1. Use engagement analysis to inform content decisions while maintaining mission priorities
2. Ask AI to create regular reporting dashboards tracking key engagement metrics
3. Have AI test hypotheses about why certain content performs well or poorly

Reflection Question
What does engagement data reveal about the gap between what you're trying to communicate and what your audience is actually hearing or needing?

CHAPTER 109
Creating Professional Development Plans for Staff

Our Goal

Design professional development plans for staff that identify growth areas, set learning goals, and create accountability for ongoing skill development.

The Problem

Staff need ongoing training and development to stay effective and engaged, but creating individualized growth plans for multiple staff members requires HR expertise and time. You want to invest in people's development without overwhelming administrative burden.

Why This Matters

Professional development keeps staff motivated, competent, and growing. Good development plans prevent stagnation, address weaknesses, and prepare people for increased responsibility. When AI helps you create these plans, you can offer quality staff development without extensive HR infrastructure.

The Prompt

```
I'm creating a professional development plan for
[staff role: associate pastor, administrative
assistant, youth director, worship leader, facilities
manager] covering the next [6 months/1 year].

Create a development plan that includes:
- Assessment of current skills and growth areas
- Specific learning goals aligned with role
responsibilities
- Resources and training opportunities (courses,
conferences, mentoring, books)
- Timeline and milestones
- How new skills will benefit ministry
- Evaluation criteria and accountability

Staff context: [experience level, current challenges,
career goals]
Church resources: [budget, time available for
training]
Development focus: [skill gaps, leadership
preparation, technical training]
```

What to Expect
AI will generate comprehensive development plans with goals, resources, and accountability measures. You'll need to customize based on individual staff members' actual needs and ministry context.

Follow-Up Prompts
"Add spiritual formation components - professional development for ministry staff should include spiritual growth."
Technical skills matter, but spiritual health is foundational for ministry effectiveness.
"Suggest affordable or free alternatives to expensive training programs."
Professional development shouldn't require unsustainable budgets.
"Create quarterly check-in questions for supervisor conversations about development progress."
Regular accountability ensures plans lead to actual growth, not just good intentions.

Your Discernment Matters: Development plans should serve both staff member's growth and church's mission. Balance individual aspirations with organizational needs.

Verify This: Ensure recommended training aligns with your church's theology and values, especially for ministry-focused development.

How to Use It
1. Use AI's development plan as starting framework and customize through conversations with staff members
2. Ask AI to create development planning templates usable across multiple staff roles
3. Have AI develop evaluation rubrics assessing whether development actually improved performance

Reflection Question
What skills or capacities, if developed in your staff, would most significantly strengthen your church's ministry effectiveness?

CHAPTER 110
Researching Best Practices for Ministry Challenges

Our Goal
Research how other churches and leaders have successfully addressed ministry challenges you're facing, learning from their experience.

The Problem
Every pastors faces ministry challenges – declining (or growing!) attendance, multigenerational worship conflicts, volunteer recruitment struggles, financial strain, or community outreach barriers. You don't want to reinvent solutions others have already figured out, but researching best practices takes time you might not have.

Why This Matters
Learning from others' successes and failures accelerates your effectiveness. Best practices research reveals creative solutions you wouldn't have imagined. When AI helps you research ministry challenges, you can benefit from the collective church experience without attending endless conferences or reading dozens of books.

The Prompt
```
I'm researching how churches have successfully
addressed [specific challenge: declining attendance
in traditional churches, engaging young adults,
building authentic community, navigating pastoral
transition, recovering from conflict, financial
sustainability in small churches, etc.].

Research best practices and provide:
- 3-5 successful approaches churches have used
- Key principles underlying effective solutions
- Case examples or stories of churches that succeeded
- Common mistakes to avoid
- Contextual factors affecting what works (size,
location, demographics)
- Resources for learning more (books, articles,
church examples)

Our context: [church size, location, demographics,
specific circumstances]
Challenge specifics: [details about your situation]
Resources available: [budget, volunteer capacity,
time]
```

What to Expect
AI will provide researched best practices with examples, principles, and contextual awareness. You'll get strategic options without exhaustive literature review.

Follow-Up Prompts
"Which of these approaches would most likely work in our specific context given our resources and constraints?"
Not all best practices transfer universally – contextual fit matters.
"What are the risks or downsides of each approach?"
No solution is perfect – understanding tradeoffs helps you choose wisely.
"Suggest churches or leaders I could contact who have successfully implemented these approaches."
Direct conversation with practitioners adds depth AI research can't provide.

⊕ **Your Discernment Matters**: Best practices from other contexts may not fit yours. Adapt principles wisely rather than copying tactics mechanically.

✝ **Theological Caution**: Some "successful" church practices compromise theological integrity for results. Evaluate approaches through your tradition's values.

⚠ **Verify This**: Research the sources behind recommendations to confirm practices and outcomes.

How to Use It
1. Use AI research to generate options and then consult with leaders who've implemented them
2. Ask AI to research multiple related challenges to understand systemic solutions
3. Have AI create implementation plans adapting best practices to your context

Reflection Question
Which of these researched approaches feels most aligned with your church's calling and values, and what would faithful implementation require?

AI Platforms Worth Exploring

The AI landscape changes rapidly, but these platforms were among the most useful and accessible for pastoral work as of early 2026. Each offers free tiers or trials, so you can experiment before committing to paid subscriptions. Start with one or two that match your immediate needs, then expand as you grow more comfortable.

1. Claude (Anthropic)   **claude.ai**
  Known for longer context windows and nuanced conversations, particularly strong for writing complex documents.

2. ChatGPT (OpenAI)   **chat.openai.com**
  Excellent for writing tasks across all ministry areas.

3. Google Gemini   **gemini.google.com**
  Integrated with Google Workspace, making it seamless for pastors and churches already using Gmail, Google Docs, and Google Drive.

4. Microsoft Copilot   **copilot.microsoft.com**
  Built into Microsoft 365, helpful for those using Word, PowerPoint, and Outlook for writing, presentations, and email management.

5. Perplexity AI   **perplexity.ai**
  Research-focused AI that cites sources, excellent for fact-checking, research, and finding scholarly resources quickly.

6. Canva   **canva.com**
  Design platform with AI features for creating graphics, social media posts, presentations, and visual content without design expertise.

7. Descript   **descript.com**
  AI-powered audio and video editing, perfect for podcast production, audio recording, and creating shareable content (ex: worship)

8. Grammarly   **grammarly.com**
  Writing assistant that checks grammar, tone, and clarity – useful for polishing everything from bulletins to pastoral letters.

9. Otter.ai   **otter.ai**
  Transcription service that turns meetings, sermons, and conversations into searchable text, great for documentation and accessibility.

10. Notion AI   **notion.so**
  Organizational workspace with AI writing features, helpful for managing sermon series, event planning, and collaborative projects.

Remember: These platforms evolve constantly. What matters most isn't mastering every tool, but learning the principles in this book that transfer across any platform – now and in the future.